The Mobile Internet Handbook

2016 U.S. RVers Edition

by

Chris Dunphy & Cherie Ve Ard

with guest author Jack Mayer

Fourth Edition: February 2016

Third Edition: February 2015
Second Edition: August 2014
First Edition: June 2013

Disclaimer: This is complicated stuff, and there are no easy, one-size-fits-all solutions. In this book, we share what we've learned over the years in our own journeys using extensive amounts of mobile internet along the way, as well as what we've learned from wide-ranging research and conversations with other mobile internet reliant RVers.

We have no formal affiliation or financial stake in any of the products or services we mention, except our own mobile apps. With anything you purchase, you are entering into transactions directly with the manufacturers and providers of those services.

This book is a sharing of our research and experience as full-time RVers ourselves. We can take no responsibility for the choices you make as a result of reading this book. We will do our best to share the pros and cons of each option as we understand them today, but ultimately you must continue your research and decide for yourself.

When issues arise, please seek support and resolution from the provider, vendor, or manufacturer you purchased from – not us!

Also by Technomadia:

No Excuses: Go Nomadic!

www.technomadia.com / excuses

A practical guide to the logistics of full-timing – income sources, mail, banking, healthcare, family, pets, safety, relationships, preparing, and much more!

The subject of this book is constantly evolving.

It's ~~almost~~ guaranteed that as soon as we submit this book for publication, there will be an industry development that makes something in this book out-dated. We're constantly staying on top of this topic at RVMobileInternet.com, a site meant to supplement this book. We post RV mobile internet relevant news as it comes out, as well as provide more in-depth guides, product reviews, and more. Join us:

www.RVMobileInternet.com

We welcome you to subscribe to our **free monthly newsletter** – we'll send you a summary of what's changed in the mobile internet landscape recently:

eepurl.com/0KJG1

Join our free public Facebook group for discussions with other RVers, ourselves included, interested in this topic:

www.facebook.com/groups/rvinternet

Premium Membership & Private Advising

We offer a premium membership service, called *Mobile Internet Aficionados*, for those who want to go deeper – with exclusive in-depth content (product reviews & guides), news analysis, webinars, and member only Q&A forums where we can help you further. If this book is the textbook, MIA is our classroom.

We also offer private advising sessions for those who'd like our help figuring out their ideal setup.

See the last chapter, The Ongoing Conversation, for more information and a money saving coupon for having bought this book already.

Dedicated to

Tim VeArd
1944 – 2013

Technology pioneer, national hero, and location-independent technology entrepreneur who inspired us in ways beyond imagination.

We miss you, Dad.

And a HUGE Thank You to all of our supporters who crowdfunded the 2014 rewrite and massive expansion of this book that made it possible for us to make this book and RVMobileInternet.com our focus!

Table of Contents

Entertainment on the Go

Crossing International Borders

Sample Setups

Wrapping Up: Top 10 Tips

Glossary of Terms

The Ongoing Conversation

Prologue to the 4th Edition

In the spring of 2013, we set out to write a comprehensive blog post about the options for keeping online while being mobile – bringing together in one place the years of content we had written on the topic on our personal blog. Our prime goal was to control the flood of questions we got on this.

By the time we were partway through, we knew this couldn't be covered in a single post – it turned out there was way more than a book's worth that could be written on the topic!

And thus, on the spur of the moment, the first edition of *The Mobile Internet Handbook* was born. We had never published a book before – and we put together the first edition in under 3 weeks on a shoestring budget, eager to get back to our *real jobs*. And it was just a handbook at 87 pages.

We did not expect the book to take on a life of its own – gathering so many reviews and endorsements as a "must read" for RVers who need to keep online. Eventually we came to realize just how needed this sort of resource was – people were hungry for unbiased information and clearly explained guidance!

Technology marches on, and by mid-2014 it was time to update the book.

To gauge just how much interest there was in a second and expanded edition, we decided to try a book pre-sale via the crowdfunding site Indiegogo.

We hoped to at least cover the upfront costs of professional graphics, illustrations, and editing so that our new book would look more like a book and less like a long blog post. We set several stretch goals for new chapters
covering frequently asked-for topics that we could also add to the content.

And in the end – we were blown away by the support we received, and every single one of our stretch goals was fully funded – and that's how our little handbook turned in the comprehensive book this is today. We really should change the title, but we've grown fond of it.

The funding also allowed us to launch a companion website which has now become a central resource center dedicated to the topic – RVMobileInternet.com. On this site, we focus on keeping this topic current – reporting on the news and keeping updated living guides.

The website also has a premium membership component allowing us to dedicate time to assisting members, hosting an interactive forum, and writing in-depth guides and product reviews.

**We are proud to be completely reader & member funded
(thank YOU!).**

We're also thrilled that Jack Mayer, a fellow mobile tech guru we've always respected, approached us to be involved with the second edition. He contributed a chapter to this book, and he has become a regular contributor to RVMobileInteret.com.

Now this brings us to this new fourth edition for 2016.

Things in the mobile internet world seem to never slow down, and we've settled into a yearly pace keeping this book updated.

With this edition, we have completely re-organized and re-written many sections - not only updating the out of date content, but revising extensively with the goal of both going deeper and making things simpler, adding all that we've learned in the last year of assisting thousands of RVers staying connected.

We have focused on making this book a primer on the theory and a snapshot of current best practices. And less on the actual products & plans.

To make sure you always have the most current information on plans, news and products – be sure to check our RVMobileInternet.com resource center for the latest.

We point out in the book what information is most likely to change and where you can find the companion living guide that is kept up to date.

We set out to create the definitive resource for RVers wanting to stay connected while on the road.

We hope that you will agree that we have.

– Chris & Cherie

Introduction

More than likely the internet plays at least some role in your life – and for many of us, it is central.

For anyone thinking of hitting the road, figuring out how to best keep *online* while exploring the world *offline* becomes critically important.

Whether you use the internet to keep in touch with loved ones, navigate to your next destination, look up accommodations, manage your finances, learn online, pursue entertainment, or depend on connectivity for your income source – building a connectivity arsenal that suits you is an essential chore most tech-connected nomads face.

Even those who don't consider themselves techno-savvy at all still face needing at least some connectivity these days.

Who This Book Is and Isn't For

This book is focused on internet connectivity options for mobile folks based in the USA.

And more specifically, some of the resources are geared specifically towards RVers and those living on the road – whether full-time or seasonal.

If you are setting out to explore the vast expanse of the USA for a prolonged period of time and want to remain connected, you've found a great resource. We also include a primer on staying connected while crossing borders into Canada, Mexico, and beyond.

We aim to present this information in a clear manner that will be accessible to most – ranging from folks who don't yet understand the difference between Wi-Fi and cellular data, to those who are super geeks like us who understand the differences intuitively.

This book isn't intended to be consumed in one sitting, nor do we expect that you'll understand it all immediately. It is meant to be a reference guide that you can refer back to when you need information on a specific component of your setup, or as your own knowledge increases.

There is no denying that at the root of it all, *The Mobile Internet Handbook* is pretty technical stuff. We try to start each topic off as fundamental as we can, and then ramp up from the basics. If you start to feel overwhelmed as it does, *that's ok*. Keep re-reading and pushing your technical boundaries.

Keep in mind, you're going to be out there on your own needing to manage whatever connectivity toolkit you assemble, often miles away from any geek help. And if you are having trouble getting online, you may end up unable to ask for assistance in forums and groups. So before you get lost in the wilds (figuratively or literally!), make sure you understand what you have and how to use it!

You may need to enlist the assistance of a geeky friend or family member to help you assemble and install your arsenal.

Don't be afraid to ask for help – before you really need it!

Who are Chris & Cherie?

We're Chris and Cherie, also known as the Technomads of Technomadia.com.

We've been traveling full time since 2006, mostly via various RV setups. Currently we call a 1961 geeked-out and solar-powered vintage bus conversion our home, but we have also full-timed in small travel trailers.

As we're both currently in our early 40s, we're too young to be traditionally retired and not fortunate enough to be retired early…yet. And we have some serious wanderlust.

Introduction

Thankfully mobile technology enabled us to take our high-tech careers in software development, strategy advising, and technology consulting on the road – allowing us to work remotely for our clients while also creating our own products.

Our business is Two Steps Beyond LLC (www.twostepsbeyond.com).

We consider ourselves to be "technomads" and have been able to create a lifestyle that combines our careers and our desire for mobility.

Needless to say, we absolutely depend on mobile internet to keep connected to our clients, manage our projects, keep in touch with loved ones and make new friends on the road.

And well, we're just geeks who like to spend a lot of time online. Heck, we both pretty much grew up with being online as part of our teen social life in the 1980s, and we met online (on a Prius – yes, the hybrid car – forum!) shortly after Chris first hit the road in 2006.

About Chris: Before going nomadic, Chris had a career in Silicon Valley focused on mobile technology.

He started out as the founding technical editor for *boot* magazine – mastering the art of explaining complex topics for a mainstream audience.

His most recent corporate job was as Director of Competitive Analysis (aka "Chief Spy") for Palm and PalmSource – the companies behind the pioneering Palm Pilot and Treo. His job was to be intimately familiar with every mobile device and technology in existence, and he was tasked with traveling the world to dig up information to chart the future of the mobile industry.

It was not uncommon for him to be carrying dozens of mobile devices with him at a time – always raising eyebrows passing through airport security scans.

Times haven't changed much actually – we still have dozens of mobile devices on board!

About Cherie: Cherie ran a software development business from home since the mid-1990s, with a long history of working remotely for her clients.

Aside from developing custom business software, her career involved technical writing and teaching high-tech topics to nontechnical people.

When she met Chris during his first year on the road, she was accustomed to carrying smartphones able to tether her laptop to the internet. And it was essential for her to keep connected if she was to join him full-time while still running her business.

Our first year on the road together was spent in a tiny teardrop travel trailer (16') equipped with just the essentials – solar electricity and mobile internet. Luxuries like air conditioning, refrigeration, and even plumbing had to wait a year until our second travel trailer (17') was custom built for us.

After so many years on the road, we have tried a wide variety of mobile internet solutions.

We know this stuff intimately, our lives are living laboratories for mobile technology, and it is our honor to share our experience.

Follow our personal nomadic adventures:

> **Blog:** www.technomadia.com

> **Facebook:** www.facebook.com/technomadia

> **YouTube:** www.youtube.com/user/TalesFromTechnomadia

> **Instagram:** @cherie_technomadia / @chris_technomadia

About Our Guest Author, Jack Mayer

Jack Mayer has been a full-time RVer since 2000. He is a freelance writer specializing in RV-related technical topics, author of a popular RV-related website (www.jackdanmayer.com), and speaker at RV rallies. His professional background is in computer system software design, networking, and operating systems.

For the past 15 years, he has designed and implemented a variety of wired and wireless networks for RV parks, small businesses, and individuals. In addition to his work in the communications field, Jack has specialized in the design and implementation of RV electrical systems for off-grid living.

More About This Book

Just some disclosures:

- Except where specifically disclosed, we are not financially affiliated with any of the products or companies we talk about in this book, and we're not writing this book to sell you anything. We're just sharing our lessons learned, our tech backgrounds, and our years of experience living on the road while keeping connected. We strive to be as unbiased as we can.

- Yes, some of our tech equipment has been provided at no charge by the manufacturers/vendors – but this usually come with the expectations of us being reviewers or beta test sites in which we provide feedback to the companies to improve their products.

  To see all of the current gear we have in our testing lab, visit: www.rvmobileinternet.com/lab

  We get a say in how these products evolve to better meet the needs of folks like us who are truly mobile. And being able to get free or loaner gear helps us review far more products without having to charge an arm and a leg for this book to cover equipment expenses. But some stuff we do buy on our own as well – we focus on finding the best, not what we can get sent for free.

  If you know of any cool mobile technology that you think we should consider featuring in a future book update or on RVMobileInternet.com, let us know. If you are a manufacturer, let us know if you would like to send us some gear to test and review.

  Beware: You can count on our honest (sometimes brutally so) feedback!

- Much of our current tech is Apple products, though in the past we were both Windows users, and we do keep current with Windows and Android developments.

  However, despite our Apple-centric personal arsenal, this book it not platform specific – all are welcome here!

Introduction

- In addition to our consulting projects, we also write mobile travel apps for the iPhone, iPad, and Android.

  'Coverage?' (www.twostepsbeyond.com/apps/coverage) is the app most related to this book, and you'll see screen captures in the book used to compare the various carriers. This powerfully simple iOS app uniquely displays overlayable coverage maps from the major carriers to help us technomads know where we can best keep connected. While we'd love to create an Android version, free time to do so seems to be something we just can't find.

  If you do buy this app, you *are* tossing a couple extra bucks into our account – thank you!

- Please don't assume any products, prices, or plans mentioned in this book are necessarily current. They were all fact checked before submitting for publication, but this stuff changes often.

  This book isn't intended to be a price comparison guide or survey of all the specific products out there. The examples included are there to help steer you in the right direction.

  Check RVMobileInternet.com for updated guides and news – we are constantly updating information there.

- Although we do provide some technical and installation advice, please always consult with the product's vendor or manufacturer for direct support.

Armed with the information in this book, you will be much better equipped to understand and evaluate the current offerings on the market – and to decide what plans and technologies personally fit you best.

Our goal here is not to give you a shopping list for the singular perfect mobile internet setup – it is to arm you with the information you need to write your own.

Laying the Groundwork

First off – if at any point you come across terms in this book that are at all unfamiliar to you, stop and check the glossary at the end!

We have written a comprehensive glossary that defines even the most technical terms in ways that most should understand.

So before you get frustrated wondering why you might need a POE to power your CPE to get remote 802.11g when you'd really rather have more dB on your LTE – check the glossary, and soon it will make better sense.

Mobile Internet Options

Basic Differences Between the Common Options

There are multiple ways to access the internet while on the go these days, and each of them has attributes that might make one more attractive than the others.

Here's a quick grid comparing the primary options:

	Unlimited or Capped	Mobile Friendly	Cost	Speed	Reliability
Cable/DSL	(Usually) Unlimited	Not mobile. Fixed location.	Reasonable	Fast!	Always on
Cellular	Usually Capped	**Fully mobile – wherever there's signal!**	Pricey	Slow to fast (Faster all the time!)	Varies by location
Public Wi-Fi	Variable	You hunt signal at each location	**Free to cheap**	Highly variable	Highly variable
Satellite	Capped	**Fully mobile – wherever there's southern sky**	Pricey to beyond pricey!	Slow and high latency	Can be persnickety

- **Cable/DSL:** This might be what you're used to 'at home' – but on the road, this option is only sometimes found in long term RV parks.

- **Cellular:** Cellular data is quite prevalent and has gotten amazingly fast – and is now even available in sometimes surprisingly remote locations. However, cellular is typically priced by how much you use, which can add up fast. You might also need extra equipment and boosting gear to optimize utilizing cellular in remote locations. You'll need to select your carrier(s) and equipment wisely to best match your planned travel destinations and routes.

- **Wi-Fi:** Public Wi-Fi hotspots are often free or low cost, but they can vary vastly in quality and are frequently too overloaded to be reliable. Unless you're willing to take your laptop closer to the physical hotspot, you may need additional gear to get a usable signal from the comfort of your RV.

- **Satellite:** Satellite can be picked up anywhere with access to the southern sky, even remote locations and across borders into Mexico and Canada – but satellite internet comes with a host of drawbacks, including speed (or lack of it), price, latency, and complexity.

More than likely, most RVers will create a personal arsenal that combines multiple options to best fit their own unique needs.

Understanding the Difference Between Cellular & Wi-Fi

By far the two most common ways that RVers will be able to get online is via either a cellular data connection or via public Wi-Fi networks, such as those offered at many campgrounds.

But – one of the most basic questions we get asked – just what exactly is the difference between Wi-Fi and Cellular?

We'll be going into much more detail later, but here is the basic breakdown:

Wi-Fi – This is a short-range local wireless network technology.

The Wi-Fi "hotspot" is a wireless access point which shares its upstream internet connection (such as cable, DSL, satellite, or even cellular) via a wireless signal that can generally be received only a few hundred feet away.

All modern laptops, smartphones, tablets and many other internet connected devices have Wi-Fi receiving ability.

Many Wi-Fi connections tend to be free – offered by campgrounds, cafes, stores, libraries and hotels. But there are some paid options out there. A Wi-Fi hotspot may also also be one you create and host yourself.

Cellular Data – This is a longer range data connection that uses the same basic wireless network that cellphones use for voice and texting – with service provided by a cellular company over licensed airwaves.

All smartphones, some tablets, some newer cars, and a very few laptops have cellular data receiving capabilities built in.

When using cellular data you are accessing the net via a cell tower that might be right next door, or perhaps as far as 20 miles away.

Cellular data is rarely free, and access requires you have a data plan with a cellular carrier such as Verizon, AT&T, T-Mobile, or Sprint.

Mobile Hotspots – Merging cellular and Wi-Fi are mobile hotspots, which are cellular receiving devices that also create a small Wi-Fi hotspot to share the cellular data connection with other devices.

A smartphone or tablet can usually create a personal hotspot that functions like this to share its connection, or a small dedicated device called a MiFi, Mobile Hotspot, or Jetpack (they're all the same thing) can do the job too.

The radios involved and technologies underlying Wi-Fi and cellular data are different – a cellular booster is not going to help you pull in a Wi-Fi signal from farther away, and extended-range Wi-Fi equipment will do nothing to improve your cellular reception.

And needless to say – cellular and Wi-Fi radios are completely different and not compatible with TV antennas either.

How Much Data Do You Need?

Mobile data is often capped or metered, not unlimited like you might be used to. Getting a handle on your actual expected usage is critical when building your mobile arsenal.

To start with, you should do an assessment of your usage and monitor it for a while – even before you hit the road.

If your internet provider doesn't provide a monthly usage number for you, it is recommended you install a usage counter on your computers and/or router to record your monthly usage (see the "Managing Mobile Data Usage" chapter later in the book for more information).

Track your actual regular usage for a reasonable amount of time (at least a full week or, better yet, a month) to get a baseline. Remember to factor in all of your devices that you'll be connecting on the road – the tablets, the smartphones, the music players, the game systems, the eReaders, and all the laptops and computers that will be in your household.

Also consider how you currently consume media content – like movies and TV shows. If you're doing that over cable TV now, how will that translate for you once you no longer have cable?

Now, compare your own personal baseline to what it would cost to buy that much mobile data.

You may be shocked by these numbers, and be realizing at this point that you're going to have to do some serious usage trimming to make mobile surfing affordable. That – or else seek out unlimited data options.

What Exactly Is Data? – Byte Sizing

Cellular and satellite service is often sold in "buckets" of a certain amount of data, often expiring within a month whether you use it or not. This data is measured by the gigabyte or megabyte.

But how much stuff does a single gigabyte equate to?

Laying the Groundwork

Everything encoded in digital format for storage in a computer or transmission over a network is made up of "bits" – literally, zeros and ones. Eight bits make a byte, and a byte is generally a single character of text.

It takes 1024 bytes to make a kilobyte (KB), and 1024 kilobytes to make a megabyte (MB), and 1024 megabytes to make a gigabyte (GB).

Text takes very little space – a book like this one (stripping away images) is only a few hundred thousand characters, less than 400KB. Mathematical data compression techniques work incredibly well on textual data – making text take up even less space in practice.

Pictures and music, however, start to move the needle a bit.

A typical consumer-grade digital camera or smartphone takes photos that average around 2.5MB in size. In other words – when it comes to data, pictures are actually worth *several* thousand words – especially if you don't reduce their size before emailing or posting them!

The modern web has grown very graphically rich – and web pages can easily consume 1MB to 5MB per page viewed, and sometimes lots more.

Streaming music online can consume 30MB to 90MB or more *per hour*, depending on the quality.

But if you really want to burn through data – video is the ultimate data hog. Especially with high-definition (HD) video, data amounts quickly begin to be measured in "gigs."

A 90-minute movie streamed in HD can easily consume 4-5 GB of data!

Uploading & Downloading

Data transfers on the internet are a two-way street, and the connection is metered in both directions.

Uploading or "upstream" connections is the data that is transferred from your computer to another computer on the internet. This might be a small bit of data – such as submitting a search to Google or posting a status message on RVillage. Or you may be uploading a large file – such as posting a video to YouTube.

Downloading or "downstream" connections refer to data that is transferred from another computer to yours – for example, reading a blog and downloading text and images, viewing a video on Netflix or YouTube, reading your Facebook home feed, or downloading a computer operating system update.

Video and audio chat applications like FaceTime and Skype use substantial amounts of both upstream and downstream data simultaneously.

What Isn't Considered Data Usage?

There are plenty of things you can do on a computer or smartphone that will not count against your monthly data usage.

If you're not currently connected, then you're obviously not consuming any internet data. And even when you are connected, anytime you're viewing files that are already stored on your computer, tablet, or phone – you don't have to worry about data usage.

Viewing files you have stored locally doesn't use up data, nor does sharing files between computers on your local network – only transferring them via the internet does.

Viewing photos that you copied from your camera or phone to your computer? No data usage – unless you decide to share them on Facebook, Instagram, or upload them online to a service like Flickr or SmugMug.

But if you use a cloud syncing service (such as iCloud Photo Library or Google Photos) to share your photos between devices, that can use substantial amounts of data copying ALL your photos to the cloud and back.

Reading an ebook that you already downloaded? Usually no data usage is involved once the book has been downloaded the first time.

And if you make cellular phone calls using your carrier's regular voice service, the call will not count against your data usage. However, if you use an internet-based service like Skype, Google Hangouts or FaceTime to have an audio conversation – that will count against your data usage.

Benchmarks for Common Internet Tasks

To help you better understand how quickly various internet tasks can burn through data, we've taken some measurements of how much data typical tasks can consume.

Actual consumption can and will vary a lot – these are just some rough examples based upon a few rounds of real world testing.

Common RVer Tasks:

Plan Your Route – Using RVillage.com and campendium.com to scout out future potential campsites, and then planning a route in maps.google.com, including checking out the satellite view of the destination to make sure the spot looks like a good fit, and finally scoping out fuel prices along the way at gasbuddy.com: **17.6MB**

Pay Bills – Sync transactions to Quicken, check balances, and schedule two credit card payments online at two different online banks, transfer funds between accounts, and check in on investments: **13.7MB**

Check in on RVillage – Visit RVillage.com, update present location, explore the map to see who else is around, and make a post to the news feed saying hello: **9.8MB**

Post a photo to Instagram – Posted a photo of us taking a measurement of data consumption of posting a photo: **2.6MB**

Send an email to Mom – A loving text email used just 70KB, with a "large" picture attached it was 10x larger: **747KB**

iMessage Chat – Short back and forth text chat, including a photo and a contact transfer: **500KB**

Tasks by Time:

To make it easier to understand how data usage adds up, we performed some everyday tasks for set periods of time and measured the data consumed. We tested five minute at a time, and multiplied by 12 to give a per-hour rate.

Your mileage will vary greatly – but this gives a good ballpark figure.

As you will see – a 5GB (5,000MB) data plan can be used up incredibly quickly, with just an hour or two of average web surfing per day likely to use up the entire allotment within a month.

Multiply by two or three to allow for multiple users sharing a connection, and it goes even quicker.

And if you throw HD video into the mix – your monthly bucket can end up gone in no time!

Be Aware of Activity Behind the Scenes

Regardless of what you are explicitly doing in the foreground – modern computers, smartphones, tablets, and other devices are often busy behind the scenes, burning data that you might not be aware of.

We carefully avoided any background activity in our tests – but if you conduct your own testing to better get a handle on your own usage patterns, make sure to take potential background data hogs into account!

For more information on effectively managing your data usage, refer to the "Managing Mobile Data Usage" chapter later in this book.

Laying the Groundwork

Online Task	Data Used per Hour
FaceTime Audio Call	40MB/hr
Skype Audio Call	46MB/hr
Listening to Pandora (Standard Quality)	52MB/hr
Reading / Posting in RV-Dreams.com Forum	71MB/hr
Online Gaming (Typical Usage)	75MB/hr
Reading Gmail	77MB/hr
Skype Low-Res Video Call	96MB/hr
Browsing RV Blogs	137MB/hr
Browsing Facebook	140MB/hr
Actively surfing twitter.com (including skimming links)	155MB/hr
Listening to iTunes Radio	180MB/hr
Surfing Google News, reading top stories	186MB/hr
Watching YouTube (360p Resolution)	221MB/hr
Listening to Pandora (High Quality)	300MB/hr
Watching Netflix (iPad SD Playback)	384MB/hr
FaceTime HD Video Call	408MB/hr
Watching Technomadia Video Chat Archive	516MB/hr
Watching Netflix (Desktop SD Playback)	1,104MB/hr
Watching Netflix (iPad HD Playback)	1,656MB/hr
Watching YouTube (1080p Resolution)	1,920MB/hr

Rule #1 – Reset Your Expectations!

Mobile internet has come a long way since we first hit the road in 2006 (yeah, yeah… we used to surf uphill both ways on 2G 1xRTT!).

The technology for connecting while on the go has advanced at an incredible pace – there is a vast improvement in speeds and coverage available today, and it's only getting better.

The price per GB of data has plummeted too, though typical monthly usage has gone up even faster than prices have gone down – so things certainly do not feel any cheaper!

Public Wi-Fi spots are all over now, and there have been so many improvements in cellular coverage that you can now get abundant connectivity across the bulk of the nation.

And sometimes, you can even get connection speeds while on the go that exceed what you could get via the fastest fixed-place connection.

> *But despite all these advancements, there are still limitations and plenty of frustrations.*
>
> *The most important thing you can do to prepare yourself for relying on mobile internet is to reset your expectations.*
>
> *Be ready for the bad days.*
>
> *The slow days.*
>
> *The no days.*

We're not trying to scare you away, but we do want to make sure your expectations are realistic. Keeping online most of the time while traveling is entirely possible – but it's not necessarily easy or cheap.

There is a scene in the film *RV* where the late great Robin Williams is standing on top of his rig like the Statue of Liberty, trying desperately to send an email only to have his battery die just as his dozenth attempt looks likely to complete.

It is an especially funny bit for us RVers... because many of us have been in that exact situation (and pose) way too many times.

The real secret to connectivity on the road is learning to be flexible and embracing rather than struggling against the constantly changing ebbs and flows of bandwidth.

Living as a technomad, some days you will have a connection that seems as if you are plugged directly into the heart of the internet, and other days you will be wishing for an upgrade to IP over carrier pigeon (en.wikipedia.org/wiki/IP_over_Avian_Carriers).

Navigating mobile internet is not going to be anywhere near as easy as just plugging in a cable like one you might have had in your fixed home. And sorry to say, they don't even make a cable long enough to take your wired internet with you across country (we tried, it got tangled up in our axles).

You will be battling:

- Intermittent and variable connections.

- Varying speeds – from frustratingly slow to blazing fast.

- Usage cap limitations on how much data you use.

If your mobile livelihood absolutely depends on keeping connected, you will have to carefully plan your mobile life around this need.

This could mean altering routing and carefully planning where to stop to get work done.

It may mean having to move on sooner than you're ready to search for better bandwidth, and may even mean afternoons spent at the local Starbucks, McDonald's, brewery, or library to soak up some Wi-Fi.

Reset Your Expectations

If you can turn the inevitable frustration around – from "Gah! I have to go find Wi-Fi" to "Oh darn, I have to indulge in a local craft beer while I get some work done" – you'll be better able to thrive with this lifestyle.

It may mean that some days you will find yourself actually thankful to be paying overage fees when a single source of mobile data is your only option in a specific location.

No matter how many backup plans you build into your connectivity arsenal, when you absolutely need to get online, that's inevitably when the glitches will emerge.

No matter how much time and money you invest in ensuring great connectivity, there will be times that you can't get a stable connection to do everything you need from the location you want to be at.

Mobile equipment can (and will) fail, firmware patches and upgrades can cause unintended problems, weather can interfere, or your exact parking location can influence your signal. And even something as invisible as your neighbor's microwave oven can conflict with your Wi-Fi network, knocking you offline until their popcorn is ready.

You need to realistically set your expectations, as well as the expectations of the people depending on you being online – such as clients, co-workers, family, and friends.

> There will inevitably be compromises in connectivity in exchange for your mobility.
>
> *What tradeoffs are you willing to make?*

Very seriously consider the costs of assembling your mobile internet options. Decide for yourself how much it is worth spending to try to cover your bases and how much internet access *you really need.*

Even if your income source does not depend on internet access, you may need to adjust your expectations around how much you rely on connectivity for personal reasons – such as email, Facebooking, viewing video content, banking, bill paying, looking up information about your next destination, online learning, gaming, and keeping in touch with loved ones.

If you're used to streaming TV and movies over your fast unlimited cable internet, you may need to adjust your viewing habits to include other sources of entertainment.

Reset Your Expectations

If online gaming is your entertainment of choice, you may need to resign from your Warcraft clan and focus on single-player or turn-based games instead.

And if you've gotten hooked on video chatting, you may need to actually resort to old-school voice phone calls every so often.

No matter what you do, there will be days that staying connected is more of a headache than it is worth. The most important rule for staying connected on the road is that you need to be mentally prepared for these days.

Our solution? We consider a box of wine an essential component of our mobile internet arsenal.

Assembling Your Arsenal

There is currently no one single technology for keeping online that is appropriate for all the different situations mobile users might find themselves in.

The most fundamental key to success for staying connected while on the road is having multiple pipelines ready to try at each location.

When Plan A is out of range or overloaded, Plan B suffers a hardware failure, a tree is blocking the signal to Plan C, and you ran over the wire to Plan D – what will you try next? How much redundancy do you need?

Each individual solution has its pros and cons, and will be more ideal in some locations than in others. Sometimes what works best in a given location even changes based on the time of day, or the weather.

Your ideal arsenal is going to be very personalized to you and dependent upon several factors including:

- **How often do you need a reliable connection?**

  Do you need to be online all the time? Or can you be offline for a few hours? How about a couple days or (gasp!) weeks at a time?

- **How much internet data do you need?**

  Do you just need to send a photo of today's great view to the grandkids, schedule a banking payment, update your location on RVillage, and do a little surfing?

  Or do you need to regularly transfer big files, VPN into remote servers, watch nonstop cat videos, host frequent video conference calls, trade stocks online, or attend lots of online video classes?

How much of this is really a *need* versus a *luxury* for you?

- **How many devices do you need to keep online?**

  Keeping one laptop online is different from keeping a large multi-device family connected.

- **What is your style of travel?**

  Are you planning to mainly boondock in remote locations far away from civilization where few cell towers exist?

  Or are you moving between commercial RV parks primarily in urban areas?

  Or will you have some combination of styles, including some public campgrounds and courtesy parking with friends?

  Will you be traveling within the USA, or will you be crossing borders into Mexico, Canada, or heading further abroad?

  How quickly will you be changing locations? Will you be constantly on the go and want a more passive setup at each stop, or will you be sticking around a while where it's worthwhile to tweak for optimal performance?

- **What plans and devices do you have now?**

  Do you have existing contracts to contend with? Do you have a coveted grandfathered-in unlimited data plan that you'd like to keep? What gadgets do you have?

- **How much complexity can you handle?**

  Some golfers make do with just a putter, and others have a caddy hauling around a dozen specialized clubs appropriate for every possible situation.

  The same is true for mobile internet – having more tools on board increases your options, but also increases the cost, complexity, and expertise required. Unless you have the time and patience to geek-out full time, simplicity may be a better route for many.

• **Your budget?**

The cost of staying connected can add up quickly, between equipment and monthly fees. Can you afford to keep multiple redundant backup options activated, even if you won't be able to use them at some locations?

Free and cheap options will have trade-offs for convenience. And even expensive options come with potential frustrations.

If you're not hitting the road in the next couple of months, please don't jump into buying all your equipment right away!

Technology changes so quickly that you are best off leaving the final assembly of your connectivity arsenal until much closer to actually hitting the road. We recommend no sooner than 6 months before you heading out, but 2-3 months is even better.

There may be new options to consider, and what you buy now may become obsolete by the time you need it.

We personally tend to go through a total connectivity refresh at least every year just to keep up, though not everyone needs to be as focused on staying on the bleeding edge (after all, our job is to update this book!).

Our current arsenal, which we detail in the 'Sample Setups' chapter at the end of the book, may be overkill for your needs, and for some it may not be robust enough. Throughout this book, we break down the trade-offs so you can decide for yourself how much of this technology you personally actually need.

Also – don't feel you have to have your total connectivity solution built on day one. Yes, some things will be easier to install when you have access to ladders and tools and trusted installers. But you can always change things up later if you find something isn't working for you, or you decide to change up your style of travel.

And never forget that trade-offs will be inevitable. It may be frustrating in the moment, but the occasional bout of disconnectedness is a small price to pay for all the incredible perks of a mobile lifestyle!

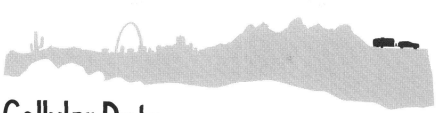

Cellular Data

Cellular data is probably the easiest and most accessible option for getting online in most places across the USA. This is where some of the greatest advancements in speed, coverage, and reliability have happened over recent years – and technology here continues to advance steadily.

Cellular data allows you access to the internet anywhere your devices can get a cellular signal from your carrier(s).

Each cell phone company (aka carrier) builds cell towers in locations it has customers to serve, and each tower transmits a signal of varying power that can be picked up by devices within its coverage zone, called a cell.

Cellular Internet

Quick Glance

Pros

Widely available

Easily accessible

Can be blazing fast!

Cons

Expensive

Data usage caps

Variable signals

Cellular Data

Those individual cells may range in size from the size of a city block to the size of an entire town (or larger!), and they are designed to overlap, creating a network of coverage.

As you move out of one cell, the connection to your device is handed off to the next tower – usually fairly seamlessly. The places where there are gaps between the cells where no tower reaches are known as dead zones, and though carriers try to avoid having any, they are sometimes inevitable.

The farther you move away from the tower in the center of a cell, the worse your signal may get – resulting in dropped calls and slower data. In addition to distance, the signal can also be impacted by terrain, buildings, obstructions, and radio interference.

The quality of your connection will also be affected by how many other customers are utilizing the same tower you are connected to. A given cell can only handle so many connections at once. To increase capacity when an area gets overloaded, the carriers build more towers, breaking larger cells into multiple smaller ones.

More than likely you're already carrying a cellular-equipped mobile internet device – such as a smartphone or tablet.

But as simple as it can seem on paper, cellular is also a sometimes a confusing subject – primarily because there are just so many options! You have to choose which carrier(s) you want, which plans make sense, what equipment to purchase, and how much speed and data you actually need.

The biggest downsides to cellular data are:

- You are dependent on where your carrier has built towers and/or has roaming partnership agreements. No one has coverage everywhere, and even if you bought a plan and device from each of the major carriers, *you would still* encounter times you just can't find a signal.

- Usage caps and network optimization practices are often in place to limit how much data you can use per month.

  While there are unlimited cellular data plans available, most of them actually have substantial limitations lurking in the fine print. Such limits include being subject to network prioritization (meaning your speeds can be slowed down when on overloaded towers), or limitations on how data can be shared with other devices.

Nationwide Carriers

The first choice to make is which carrier, or carriers, to get service with to best cover your mobile data needs. In the US, the four major nationwide carriers are (ranked by size):

- **Verizon**
- **AT&T**
- **T-Mobile**
- **Sprint**

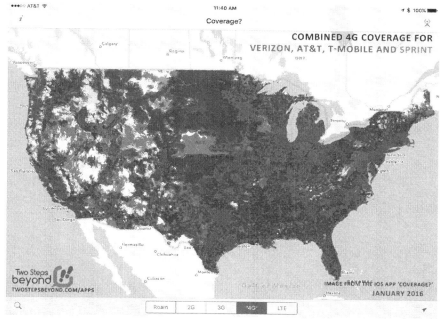

All four have embraced the same underlying fourth-generation (4G) cellular network technology, known as LTE. But they all have very different legacy 2G and 3G networks, coverage maps, compatible devices, supported frequencies, and expansion plans going forward.

If you were living stationary in one city or neighborhood, you could ask friends for their experiences with their carriers to determine which would serve you best in the local area.

But as a traveler, you will be moving around a lot – and in different locations, different carriers excel. You need to pick a carrier that is well suited not just to your home turf, but also for all the places you plan to go.

> The most important point we want to highlight here is that there is no singular best carrier for every nomadic traveler!

Cellular Data

Verizon

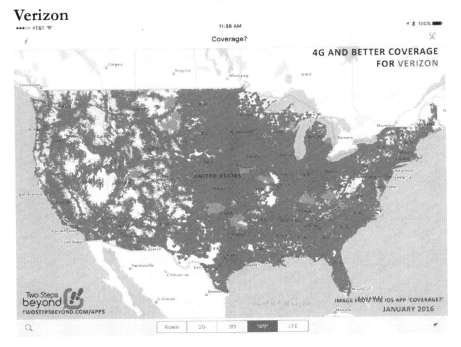

Verizon is the largest cellular carrier in the USA. It has the widest overall coverage area, the most deployed LTE, and great overall performance. Verizon has a promising future roadmap too – with an ever-expanding network and advanced XLTE service widely deployed.

Verizon now offers some video streaming via its go90 mobile app that doesn't count against your data usage, and you can use many Verizon data plans while traveling internationally for an additional $2-10/day.

For all these reasons, if you're only going to choose one network – Verizon is the natural top choice. But current Verizon plans are also pricey.

And because Verizon's network is known to have the widest coverage and has become a top choice for RVers, it's actually not uncommon to pull into a popular RVing area or large rally to find the local Verizon tower overloaded and sluggish during peak times.

Verizon does not offer an unlimited data plan to new customers, but it continues to honor grandfathered unlimited plans. These are completely unthrottled, with no usage caps, and can be used as a mobile hotspot.

At the moment – you can still obtain these plans by jumping through some hoops, but the process can be daunting.

We overview these grandfathered unlimited plans further in the 'Unlimited Data Options' chapter, and we consider this a worthwhile option for most any RVer who consistently needs at least 15-20GB of cellular data a month.

AT&T

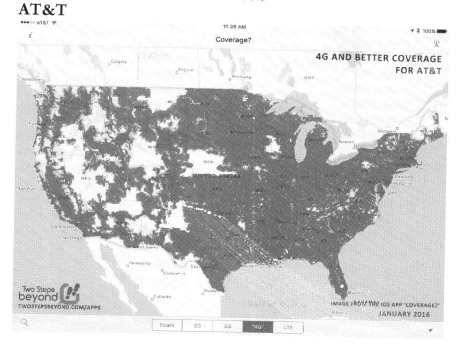

AT&T is the second-largest carrier and is a formidable rival and a great complement to Verizon for us nomads.

AT&T's LTE network lags Verizon in coverage and speed, but AT&T's older 3G and 4G networks have great coverage and speeds that blow Verizon's legacy 3G out of the water.

There are plenty of locations we have been where the best Verizon can manage is 3G or very weak or overloaded LTE. Sometimes in those locations AT&T ends up delivering a much better overall experience.

Some key features for AT&T include 'rollover data' – if you don't use up all of your data one month, it rolls over to be used the next month.

AT&T also includes free coverage in Mexico with 1GB of data a month.

A combination of Verizon and AT&T on board gives the widest coverage across the country – and we personally rely on this combination to maximize our connectivity options.

AT&T now owns DirecTV, and offers new unlimited smartphone data plans only to DirecTV customers. AT&T continues to honor older grandfathered plans. However, AT&T's unlimited plans are on-device only (no tethering/hotspot usage allowed) and are subject to network de-prioritization after 22GB of usage in a month. In other words, AT&T's unlimited plans are not overly useful as a primary mobile internet solutions. And AT&T's tiered data plans are on the pricey side too.

T-Mobile

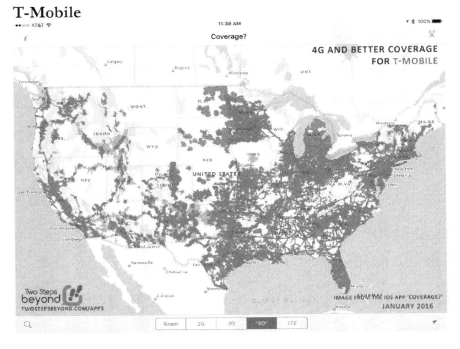

T-Mobile has been the carrier to watch this past year – blowing past Sprint to take a solid third place in the market, and doubling its LTE coverage area along the way. But in our experience T-Mobile is still a distant third behind Verizon and AT&T, especially outside of major metro areas.

T-Mobile offers some unique "uncarrier" features that make it a very compelling choice for RVers. These features include: Data Stash (unused data rolls over for use within a year), Binge On (free lower resolution video streaming from many partner providers), Mobile without Borders (free usage of your plan, at high speed, in Mexico & Canada), unlimited international data (at slow speeds) and free unlimited music streaming.

While T-Mobile has the most rapidly expanding network, only the newest T-Mobile devices are compatible with LTE Band 12 and can connect to the "extended range LTE" signal. And especially without a Band 12 device – in a lot of the more rural areas where RVers frequent, T-Mobile disappoints.

But when you have a solid T-Mobile 4G/LTE signal – its network speeds are consistently some of the fastest.

T-Mobile does offer an unlimited data plan on smartphones. It includes a generous 14GB of tetherable high speed data to share with other devices, and is subject to network de-prioritization after 25GB of usage. Since T-Mobile voice roams onto AT&T towers, this can make T-Mobile a very interesting secondary network choice for RVers. Picking up a $35/month T-Mobile tablet plan for Binge On video streaming is a great way to go too.

Cellular Data

Sprint

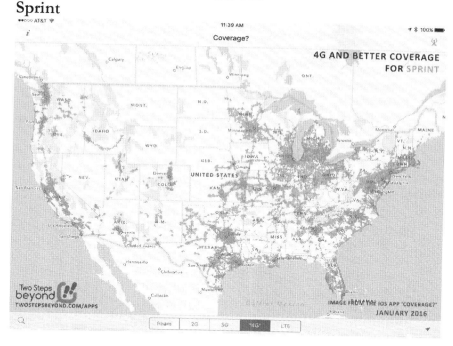

Sprint's biggest advantage is that it tends to have the most affordable options, especially the promotional plans offered to lure new customers.

But the biggest downside of Sprint for nomads is the limited coverage map.

The vast bulk of Sprint's usable fast data coverage is pretty much only found in core urban areas and along major interstates. Anything outside of that, and you're roaming with very slow speeds – if you can get online at all.

If you're planning to stick to urban areas, Sprint might be worthwhile – and Sprint's tri-band LTE Plus network shows great promise for the future.

However, Sprint has been struggling – and in 2016 it will be trying to reduce its operating expenses by relocating equipment to less expensive towers. If Sprint follows through with this plan, it is predicted to cause potential disruptions in service, and shifted coverage maps as the service transitions.

Sprint does offer an unlimited smartphone data plan, however it only includes 3GB of tethering/hotspot data to share with other devices – and is subject to slowed speeds on congested networks after 23 GB of usage a month.

There are occasionally Sprint resellers who offer truly unlimited Sprint data plans – but these offers seem to come and go frequently.

Keep Current Alert:

The carriers are constantly changing up their offerings and features to stay competitive with each other.

Before choosing your carriers, be sure to check the RV Mobile Internet Resource Center for our current take on each.

We keep this free article updated with our analysis of the carrier and the options available for each:

The Four Major US Carriers – Which is Best for RVers? (http://www.rvmobileinternet.com/four-carriers)

We keep this premium member-only shopping guide updated throughout the year to cross compare current data costs for all of the carriers and their resellers:

Cellular Carrier Data Pricing & Plan Guide (http://www.rvmobileinternet.com/resources/cellular-carrier-data-comparisons-pricing-plan-guide)

Which to Choose?

For nomads who need the widest amount of cellular coverage, we generally find that combining Verizon and AT&T gives the most options. There are simply too many locations where one carrier excels over the other.

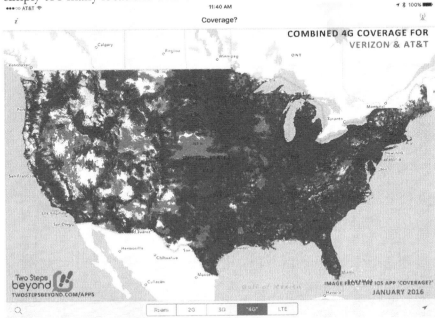

Verizon combined with T-Mobile has also become a very solid choice too.

Cellular Data

For nomads seeking to have only one carrier, Verizon is usually the top choice – but AT&T's coverage map could also be considered.

For most, we generally don't recommend T-Mobile as a primary carrier, but given T-Mobile's rapidly expanding network and its unique features, we find T-Mobile worth considering as part of a mobile internet arsenal as a secondary/tertiary carrier.

At the moment, we generally recommend skipping Sprint – unless you know for certain you will be traveling in strong Sprint areas.

As a reminder, even having all four networks onboard is no guarantee that you'll be able to get online everywhere.

Each carrier offers plans directly that you can obtain either on their website, in stores around the country, or via third-party branded resellers known as MVNOs (covered later in this chapter).

You can get plans that are for data only, or for data combined with smartphone or tablet service.

Regional & Local Carriers

In addition to the big four national carriers, there are a number of smaller regional and even local carriers that own and operate their own wireless networks.

Some of the larger examples of this sort include U.S. Cellular, C Spire Wireless, nTelos, Cellcom, and Cellular One.

These smaller regional carriers are usually poor choices for travelers, unless you know that you are primarily going to be spending time in areas where they have a strong presence.

Even if the regional carrier has nationwide coverage through roaming agreements, if you're utilizing the service primarily outside its home region, you can find yourself running into all sorts of imposed limitations that can even include having your service summarily terminated.

MVNOs, Resellers & Prepaid Plans

Mobile Virtual Network Operators (aka MVNOs) do not own their own cellular networks, but rather they buy service in bulk from the major carriers and resell it under their own brands with their own terms.

Via an MVNO you can get better deals and higher usage caps than you ever could by buying direct.

Usually an MVNO is not allowed to advertise who is providing the underlying network, but if you explicitly ask or do some online digging, it isn't hard to figure out which data network is behind any given plan.

Some brands are actually owned and operated by the big carriers themselves – allowing them to target smaller markets and niche services without diluting their national brand.

Most carriers also offer prepaid or pay-as-you-go plans that allow customers to only pay for service when they need it, rather than making month to month commitments. This makes setting up service super-easy without requiring signing contracts or going through credit checks.

Here are some of the reseller options we know of that include at least some mobile hotspot usage that can be shared with other devices:

Verizon Based

- **Verizon Prepaid** (www.verizonwireless.com/prepaid/prepaid-jetpack/) – Available direct from Verizon on its LTE network. Offers prepaid LTE Jetpack hotspot plans offering up to 10GB of data that expires after 2 months. They also have smartphone plans with 2-6GB of full speed data a month that can be used as mobile hotspot, with affordable data reloads.

- **StraightTalk Hotspot** (www.straighttalk.com) – Available at Walmart and online, they have a hotspot only plan with up to 7GB of data can be purchased on a pay-as-you-need-it basis with usage spread out over up to 2 months. StraightTalk's smartphone plans currently do not officially allow mobile hotspot use, so they don't tend to be ideal as an internet solution.

- **Blue Mountain Internet** (bmi.net/internet/mobile-broadband-rental.html) – Offers no-contract 3G/4G Verizon-based plans with up to 20GB of data a month. They provide "optimizer software" for Windows & Mac machines that claims to effectively triple your usage for non video files.

AT&T Based

- **GoPhone Mobile Hotspot** (www.att.com/gophone) – AT&T directly offers its own prepaid service, which is available at Walmarts around the country. Data is purchased as you need it, up to 5GB at a time, that expires within a week or month. All of the GoPhone smartphone plans include mobile hotspot usage.

- **Harbor Mobile** (www.harbormobile.com) – An AT&T reseller offering affordable smartphone and data only plans to businesses. All of their smartphone plans include up to 20GB of free hotspot use to share with other devices.

- **Cricket Wireless** (www.cricketwireless.com) – Owned by AT&T, Cricket offers 5GB and 10GB smartphone plans with mobile hotspot usage available as an extra charge.

- **StraightTalk Hotspot** (www.straighttalk.com) – Almost identical to the Verizon option, except on the AT&T network.

- **Consumer Cellular** (www.consumercellular.com) – Offering plans on AT&T, they have no-contract, build-your-own plans starting at just $10/month. Add on just the services you desire – from voice, text, and data. None of their plans include mobile hotspot use.

T-Mobile Based

- **Simple Prepaid** (http://prepaid-phones.t-mobile.com/prepaid-plans) – Direct prepaid plans with T-Mobile, offer lower priced plans for smartphones and data only plans for tablets and hotspots. Prepaid plans don't include the free video streaming Binge On capability, or some other postpaid perks however.

- **MetroPCS** (www.metropcs.com) – Owned by T-Mobile, MetroPCS offers smartphone plans that include mobile hotspot for data, including an unlimited plan that is capped at 8GB/month of hotspot usage. These plans do not include features such as Binge On unlimited video streaming, or included international roaming.

- **Unlimitedville** (www.unlimitedville.com) – In early 2016 they just introduced a T-Mobile based rental plan on a no contract basis. It's data only, unthrottled, and able to be used in Canada & Mexico.

Sprint Based

- **Unlimitedville** (www.unlimitedville.com) – A business reseller of Sprint that offers a truly unlimited data plan with with a 2 year contract. Available only to qualified businesses, and may be a limited time promotion.

- **Karma Mobility** (www.yourkarma.com) – A data-only option requiring purchasing their hotspot. You can then choose the Refuel plan and buy data as you need it that never expires. Or the Pulse plan that includes a set amount of data a month. If another user connects to your Karma hotspot, you both earn bonus data.

- **Republic Wireless** (www.republicwireless.com) – An MVNO with flat rate smartphone plans that default to Wi-Fi use for data and calls, and then falls back to cellular when needed. Data can be purchased as you need it, and unused is refunded at the end the month.

- **Boost** (www.boostmobile.com) – Offers a data hotspot plan that includes 10GB of data, and has affordable smartphone plans. Their unlimited plan includes 8GB of mobile hotspot use.

- **Virgin Mobile** (www.virginmobileusa.com) – Virgin Mobile, owned by Sprint, offers no-contract 'Data Done Right' plans for smartphones that include hotspot usage. Their single line plans have the option to add on mobile hotspot use by the day.

- **EVDODepotUSA** (www.evdodepotusa.com) – Aimed at rural customers, they offer an unlimited 3G and high-limit LTE plans.

- **FreedomPop** (www.freedompop.com) – Offers a free 500mb of data per month, with data reloads.

- **TruConnect** (www.truconnect.com) – TruConnect offers no contract plans for phones and data only devices. Their unlimited data plans only include a few GB of high-speed hotspot usage though.

Multi-Carrier MVNOs

- **Project Fi** (fi.google.com) – Google's unique MVNO operates on both the T-Mobile and Sprint networks behind the scenes, giving you a merged coverage map on one device. The catch with Project Fi is that currently only three Nexus smartphone are supported, and you need to request an invitation to even sign up. Project Fi offers an affordable flat rate for voice/text usage, plus per GB pricing for data that is tetherable and can be used overseas. You get a refund each month for any data you don't use.

The MVNO Landscape is ALWAYS Changing

There are so many more MVNO choices out there; for an overwhelming list of the options, check the list of MVNO's on Wikipedia. (en.wikipedia.org/wiki/ List_of_United_States_mobile_virtual_network_operators)

Reminder: Options listed above are current as of February 2016 – but they are constantly changing!

The Downside of MVNOs & Resellers

Keep in mind that carriers often favor their own name brand service over the cut-rate wholesale services they offer to MVNOs. And sometimes reseller plans that appear on the market are not officially authorized by the carrier.

While service provided by an MVNO may take advantage of the same underlying network, the MVNO may have speed caps in place, be de-prioritized over direct carrier traffic, or may not be able to take advantage of the underlying carrier's roaming partnerships.

And if there are network issues – you have to convince the MVNO's support staff to elevate the issue to the carrier, since MVNO customers are not entitled to direct carrier support.

But the biggest downside of MVNOs is that the offerings may not last.

Every so often a reseller comes along with an offer that seems too good to be true. And all too often – these offers, and even the MVNO itself, do not last for long once the underlying network provider starts to object or they realize their costs to deliver the service are higher than anticipated.

Remember these?

- **Millenicom:** A previous staple in many RVer's mobile internet setup for affordable Verizon data plans, until Verizon pulled the plug in 2014.

- **OmniLynx:** In 2015 RVers who jumped quickly were able to get a couple months of unlimited Verizon data cheap from OmniLynx - before that deal evaporated and service was discontinued.

- **Karma's Neverstop:** Originally launched in November 2015 as an unlimited Sprint plan, but limits were slapped in place once Karma realized it could not afford the amounts of data its customers were consuming. Soon after Karma completely discontinued the plan.

Cellular Data

The important lesson here is to be ready to jump on great offers when they come along (we announce them at RVMobileInternet.com), but to be wary.

Always have a backup plan in place in case service gets cut off suddenly.

Always keep in mind that the "no contract" terms go both ways!

Always resist temptation to break existing contracts or give up irreplaceable grandfathered plans for a reseller's offer which may not last.

Shopping for a Cellular Plan

Each carrier and MVNO will have its own set of plans, ranging from prepaid plans that are ideal for short-term needs or those who want more flexibility, to post paid plans. that generally require passing a credit check.

When comparing plans, here are some things to consider:

- None of the major carriers require contracts anymore, that practice was generally eliminated at the end of 2015. But the option still exists, so make sure you know if you enter one and what the penalty is for breaking it or suspending it. You may find your travel plans change (such as taking an extended international excursion or a nomadic hiatus) or that new technology comes out that you want to switch to. Try to minimize the number of long-term contracts you get yourself tied to.

- Can you put your plan on "vacation" or suspend it? And if so, what are the fees for doing so? This comes in real handy if you find yourself staying in a spot for a while where you don't get enough signal to keep online, or where you might have access to other reliable options.

- Is the device you're buying subsidized by the carrier, or will separate payments be required? Will you need to pay off the device should you switch carriers in the future?

- If you're shopping an MVNO or reseller, which underlying carrier is the plan you're considering built on? Does the MVNO have coverage limits, or limited roaming?

- Look for carriers and devices which support Wi-Fi Calling, which will allow you to place calls over Wi-Fi when there is no cellular signal present.

Shared Data Plans

Modern AT&T and Verizon plans offer a shared bucket of data that multiple lines tap into. With these plans, phones get unlimited voice minutes and texting. You just pay a per line access fee for each device you want to add to the plan that shares the data amount you select.

T-Mobile offers family plans in which you can connect multiple phones together on the same account and get a discount; however, each phone has its own bucket of data – which may make better sense if you don't want to manage data usage amongst your group.

Sprint tends to vacillate on what they offer, usually reacting to what the other carriers are offering. Currently they follow AT&T and Verizon's shared data model.

All of the carriers of course also offer individual plans. If you have trusted friends or family, it may still be worthwhile teaming up and getting a shared plan to reduce costs.

> **Tip:** Check back with your carrier periodically to see if they are offering new plans since you signed up, or are running promotions. If there are newer and better deals available, you can oftentimes switch, but you have to ask. For instance, in October 2014 both AT&T and Verizon offered 'double data deals' for a limited time. Many informed consumers were able to snag 30-60GB plans for half price, and keep the pricing for as long as they like.
>
> (Keep tuned to the RV Mobile Internet News center at http://www.rvmobileinternet.com/news – we report these offers as we learn about them. You can subscribe to our RSS feed, or our free monthly newsletter too.)

Knowing Where to Find Signal

Whether you go direct with one (or more) major carriers, or you get service via an MVNO – a critical part of selecting the right carrier(s) for you is knowing where they have coverage.

And once you have service, it's helpful to know where along your routes you'll get signal.

Here are some handy resources for tracking this sort of information down:

Checking the Carriers' Online Maps

If you have service via an MVNO, the MVNO may not have a coverage map online. But if you know who the underlying network provider is, you can go right to the source.

- **Verizon** – www.verizonwireless.com/wcms/consumer/4g-lte.html

- **AT&T** – www.att.com/maps/wireless-coverage.html

- **T-mobile** – www.t-mobile.com/coverage.html

- **Sprint** – coverage.sprint.com

Coverage? App

Although we can go to each carrier's maps online to scout out ahead if our next campground will have coverage, we decided to make it even easier. We wrote an app for that!

Our iOS app Coverage? (www.twostepsbeyond.com/apps/coverage) overlays our versions of the four major carriers maps, so you can create a personalized coverage map for the carriers you travel with.

The maps are stored on your device, so you don't even need to have coverage right now to find out which direction to head so you can participate in a 2 p.m. webinar.

All of the coverage maps displayed earlier in this chapter are from our app.

Crowdsourced Coverage Maps

Of course, just because a carrier claims they have coverage, that doesn't mean you'll be able to find it. There are some wonderful resources out there that aggregate crowdsourced signal and speed reports to create a coverage map based on measured reality.

Here are some of our favorites that we utilize:

- **OpenSignal** – Online at www.opensignal.com and they also have an iOS and Android app of the same name.

- **RootMetrics** – Online at www.rootmetrics.com and they also have an iOS and Android app called CoverageMap.

- **Sensorly** – Online at www.sensorly.com and they also have an iOS and Android app of the same name.

The downside of crowdsourcing is the maps available are only as useful as the data they collect from users of their apps. These resources tend to have good data for urban areas where they have a strong user base. But when you get to smaller cities and back roads, these maps commonly show no coverage at all.

These apps are a great compliment to Coverage? – first you can research where the carriers claim to have service, and then you can research user-submitted reports to get an idea on actual performance.

Campground Reviews

Since so many RVers depend on a solid internet connection, you'll frequently find reports of cellular coverage (and Wi-Fi performance) hidden within campground reviews. Here are some of our favorites:

- **Campendium** – Launched in 2015, the campendium.com review site includes specific fields for reporting coverage on each of the major carriers. This site tends to be frequented by bandwidth hungry RVers, and lists commercial, public and free camp spots.

- **Freecampsites** – freecampsites.net is a database of remote camping and boondocking options, and reviews might mention cellular coverage.

- **RVParkReviews** – This is one of the longest running review sites, and fellow RVers tend to leave coverage reports on rvparkreviews.com.

- **RVParking** – Not nearly as active, but you might find coverage reports within the reviews at RVParking.com.

- **RVParky** – Another popular review site is RVParky.com where you might find coverage mentions.

And of course, all of the RVing community benefits if you leave reviews too, including reports on your cellular signal.

Cellular Pitfall: Data Caps

Cellular data is generally dished out in monthly buckets with data caps (also called tiered data plans), setting a limit for how much data you can use before you are throttled down to a slower speed or charged overage fees.

The monthly cap on a plan might be marketed as 5GB, which allows you to use 5 gigabytes of data, totaling up your usage both sending and receiving.

Things to keep in mind with cellular data caps include:

- Even if the file you're transmitting doesn't complete, you are charged for the data used. So if you're uploading a video to YouTube and it fails halfway through – you've still used the data to load half the video, even though your mission was not completed.

- If you don't use your full data allowance by the end of your billing month, on Verizon & Sprint you lose it. AT&T and T-Mobile offer rollover data, but AT&T's has to be used the very next month.

Cellular Data

- The carriers oftentimes don't tally up your data usage in real time – it can be hours behind. This can make it very difficult to manage your data as you get close to your cap.

- If you exceed your cap, your carrier may charge you overage fees – which are usually billed in increments of 1GB – typically $10–20/GB on most current plans. But overage charges can be painfully more expensive on older plans or if you are caught roaming. And even if you only go slightly over by just a few KB in a month, you're charged for a whole extra gigabyte.

- Because so many users get upset with unexpected overage fees, some plans now offer unlimited usage with a high-speed data cap instead. When you exceed your high-speed cap for the month, rather than being charged an overage, your connection is throttled down to a snail's pace until the end of the month.

- Unlike when the carriers used to charge for cellular phone minutes on a peak versus nonpeak time basis, there's no equivalent with cellular data. No matter what time of day you use it, cellular data counts the same. However some unlimited data plans are subject to network management, and heavy users may be slowed down during peak usage times on congested towers.

But the biggest tip we can offer is that many plans allow you to change your data cap ant time during the month. So if you find you need more data or are about to go over, just increase your cap to avoid overages, and decrease it next month. Some carriers will even let you back date the change after you've exceeded your data cap to avoid overages. Just be careful if you have a promotional data plan (such as a double data deal), as you can not change the data allowance and keep the promotion.

Sharing: On-Device Data vs. Hotspot / Tethering Data

In your research, you'll probably encounter a cellular plan that says there's included data – potentially even unlimited data. Just be sure to get clarification if the plan is for use only on your smartphone or tablet, or if that data is able to be shared with other devices, such as your laptops.

Many phones and tablets today, especially smartphones, can be used as modems for your computers – allowing you to share your phone's data plan with your other devices either wirelessly or directly wired.

- **Tethering** – This is the term used when you are using a USB cable to connect your cellular device to your router or computer, sharing the connection. A nice bonus – this keeps things charged up while sharing!

- **Personal or Mobile Hotspot** – This refers to when you turn your cellular device into a Wi-Fi router creating a hotspot, allowing it to share its internet connection with other nearby devices and computers wirelessly.

Some cellular plans prohibit sharing any of your data with other devices. They are intended only for checking email and browsing from the device itself.

Other plans allow the included data to be shared however you wish.

And others place certain limits on how much shared data is included.

Thankfully, we're seeing a trend of more and more plans enabling sharing, and it is usually explicitly advertised as a perk. Look for words like 'mobile hotspot' and 'tethering' in the list of features. But even plans that allow hotspot use may have limitations associated with it, such as requiring multiple lines, or it may be an extra charge.

There are some hacks and work-arounds on some devices that enable you to bypass blocks and enable sharing data on plans that do not support it as a feature, but doing so can violate your terms of service. And some carriers are cracking down on this (we hear reports from T-Mobile and AT&T customers) to block the practice.

Know what the potential risks are, and decide if you're willing to take them, before pursuing these work-arounds.

Understanding Roaming

Roaming is when a cellular carrier has agreements with other networks to utilize their towers, helping the carrier provide connectivity to their customers who are just passing through areas they don't directly service themselves.

Roaming is essential to the regional carriers and the smaller national carriers, since they lack the vast networks of AT&T and Verizon. But even AT&T and Verizon use roaming behind the scenes to help flesh out their coverage maps.

The guest carriers pay very high fees for roaming data. T-Mobile revealed in 2014 it paid AT&T $181/GB for roaming onto AT&T towers!

Though carriers rarely charge you extra for domestic roaming – they tend to have special data roaming limits to keep your usage from costing them too much.

Cellular Data

Here are the data roaming limits for each of the four major US carriers:

- **Verizon** – None. They truly do have the most extensive network, and actually have very few roaming partners. They impose no restrictions.

- **AT&T** – After you've used 400 MB of roaming data in a month, AT&T reserves the option to suspend your roaming services for the remainder of the month, or to even impose fees.

- **Sprint** – Sprint allows 100-300MB roaming cap per month on all of their data plans. If you go over 300MB, they can start charging overage fees of 25 cents per MB. That can add up!

- **T-Mobile** – After 200MB of domestic roaming, T-Mobile suspends your roaming service, cutting you off for the remainder of the month.

Most of the carriers will send you a text message and/or email to let you know you've hit your roaming cap.

Knowing When You Are Roaming

On most mobile devices, right next to where the signal strength you are getting is displayed, you can see the name of the carrier you are connected to. Usually this is set to your provider - such as "AT&T" or "Verizon".

But, when you're roaming, you might see another carrier's name listed instead - letting you know that you are in a roaming zone.

By default, some carriers and devices don't switch the carrier's name displayed until *after* you've hit your roaming cap. If you're planning to be traveling a lot and want to be better able to tell when you are roaming, you may want to call your carrier and ask if they can make that switch now.

We actually were first inspired to create our Coverage? app after running into roaming limits. With the app, we had a way to quickly check if we were in a roaming area before paying campground fees.

Controlling Roaming

Some smartphones and cellular devices have options to disable data roaming in the device settings. Make sure to keep this setting off unless you are well aware of the potential costs and limits of roaming with your carrier. Especially near international borders - accidental roaming can be expensive!

Cellular Data Gear

Once you decide on the carrier(s) you want in your arsenal, you have to decide what specific equipment makes the most sense for getting online.

The basic options include devices that are restricted to data only — such as mobile hotspots and modems, or putting cellular connected tablets and phones to work serving double duty by providing an internet connection for your computers as well.

All of the big four national carriers have agreed on the same 4G networking standard — LTE. But despite the core standard, it is common for cellular phones and devices to be optimized to work with just a single carrier. Multi-carrier devices are still rare, though growing less so.

This is because all four carriers are building their networks on different frequency bands, and they also have older incompatible legacy networks to support 2G, 3G, and voice services.

What connectivity hardware makes the most sense for you depends on how many devices you're trying to keep connected, how many people you need to provide internet for, how many data plans you want to keep active, where you think you'll need internet access, and what devices you already own.

Regardless of which method you decide on, for maximum coverage and speed we recommend purchasing the newest cellular devices you can, and plan on replacing your hardware as often as every year or two to stay current.

All of the carriers are expanding their networks, and newer equipment is what gets you access to latest frequencies and bands. Be sure to also check the "Understanding Cellular Frequencies" chapter to understand the bands each carrier utilizes, and match them up to the equipment you purchase.

Cellular Data Gear

When it comes time to shop for your specific equipment, you can check our current reviews and comparative guides found at RVMobileIntenet.com.

We have member guides comparing the current options with links to reviews we've done:

Cellular Modem & Hotspot Guide:

www.rvmobileinternet.com/resources/lte-mobile-hotspot-modems/

Mobile Routers:

www.rvmobileinternet.com/resources/cellular-wifi-as-wan-routers/

Mobile Hotspots (Jetpacks & MiFis)

Also sometimes referred to as a Jetpack or MiFi, mobile hotspots are small self-contained units that receive a cellular data signal and then broadcast a Wi-Fi hotspot that enables your other devices to get online. They are a cellular modem and Wi-Fi router combined.

Most mobile hotspots tend to be able to serve 5–15 devices at once. They have a battery built into the device, which allows you to take it with you when on the go, and some even allow you to charge other devices off of them. They can also usually be directly tethered via USB into a cellular aware router, like those from WiFiRanger, Pepwave, or Cradlepoint.

Advantages:

- You can take your internet with you with these self contained gizmos.

- When working optimally, these can be a fairly easy plug-and-play solution ideal for users who don't want to have to learn to manage other more complex options.

- It's a dedicated device that can be left in your tech cabinet, plugged in, sitting next to a cellular booster antenna, and mostly forgotten about.

- Most of the newest cellular technology seems to be released on this style of device rather quickly, so this is one of the easiest and most affordable ways to update a mobile internet arsenal and to stay current.

Disadvantages:

- They have a good amount of complex software installed inside them to allow them to function as a router and create a hotspot. Newer ones seem to be better, but some of the older models were quite buggy.

Recommended for: Multi-person or multi-device households, those who depend on cellular data for critical tasks, those who want access to the latest network technology, those who don't want to fiddle with tethering from a smartphone, those who need their RV to have internet access even while they're away - such as remoting in to check on home automation systems.

USB Modems

A USB stick modem that needs to plug into something in order to be functional – either your computer or a compatible cellular-aware router.

Advantages:

- Because of this simplicity, dedicated modems tend to be more reliable than mobile hotspots. There is simply less that can go wrong. There's not much firmware on them to control functionality, that is left to the device you plug into.

Disadvantages:

- Only certain routers support connecting via USB cellular modems. WiFiRanger, CradlePoint, and Pepwave are the brands to look for with the broadest range of support.

Recommended for: Solo travelers needing just one laptop connected, or a household planning a cellular optimized router anyway.

Smartphone / Tablet Connection Sharing

Most smartphones and cellular enabled tablets can create a Wi-Fi hotspot and/or be directly tethered via USB to share their connection.

The direct carrier plans include the feature on their tiered data plans, but you'll need to look closely at unlimited, prepaid, reseller, and MVNO plans to see if sharing the data connection is supported.

Cellular Data Gear

Each device will be different in how you turn this feature on, but for many, it's just a switch you turn on in the device's settings.

On iOS it is called "Personal Hotspot" – and you can easily configure a network name and password to protect the connection.

Advantages:

- No extra device fees on your cellular phone account, on mobile share plans, tethering is included at no extra fee.

- No extra devices to manage.

Disadvantages:

- Not ideal for multiple people in a household – what happens if the person with the hotspot-enabled smartphone takes it with him or her to run errands?

- Talking on your phone can sometimes take your devices offline or greatly reduce the network data-connection speed. So if you need to regularly talk on your phone AND be online, this may not be ideal.

- Many devices go to sleep when there's no active connection going on, so you may need to wake your device up after a period of inactivity to get back online.

- If you have other Wi-Fi routers in the household, the smartphone/tablet can get confused when trying to figure it out if it should connect to the router or be transmitting a Wi-Fi signal itself.

- Hotspotting drains the battery pretty quickly – make sure the device is plugged in while sharing the connection.

- If you have any need to remote in to your RV while you're gone, usually you take your smartphone with you – thus taking the internet with you.

Recommended for: Solo travelers, for those not dependent on internet for critical tasks, for access to a secondary cellular network (ie. if your primary is Verizon with a MiFi, perhaps you access AT&T when needed from a smartphone) or for 'out and about' internet access away from the RV.

Cellular Integrated Routers

And last, but not least – there are higher end options for sharing a cellular data plan.

- **Cellular Integrated Home Router:** While MiFis are designed to be pocket sized and battery powered, there are heftier routers available from some carriers designed for residential home use. One example is the Verizon LTE Broadband Router with Voice, which is typically marketed as a solution for providing both home internet and phone service. But some RVers have also put it to good use on the road. It gives a home phone line that works with regular cordless phone sets so you can have a front and a bedroom handset, as well as a Wi-Fi network and even wired ethernet ports, all provided via a data connection through Verizon's LTE network.

- **Commercial-Grade Cellular Integrated Router:** If your goal is building a bulletproof mobile office, rather than cobbling together pocket-sized personal hotspots or repurposing home routers, you can instead invest in a commercial-grade router with a closely integrated cellular modem paired directly to it.

This is certainly a pricier alternative – but for some mobile professionals, the increased reliability may be an essential business expense.

Other higher-end commercial-grade products from Cradlepoint and Pepwave go even further in capabilities, focusing on both speed and reliability.

Disadvantages: Cellular-integrated routers tend to be updated infrequently, so they often lag a year behind mobile hotspots and smartphones in supporting the latest cellular frequency bands and technologies.

We have an entire chapter later in the book going deeper into routers.

Smartphone Selection Tips

A cell phone used to be for one thing – making calls. Today, however, making calls is just another app on what has become an incredible pocket-sized technological Swiss Army knife.

For many people, a smartphone has replaced their camera, their GPS navigation system, their game console, their CD rack, and for some it has replaced their need for a separate computer entirely.

And for many, smartphones have become a primary conduit to the internet – both directly and as a hotspot serving other devices.

Choosing a smartphone is a personal decision – and there is no universal best pick. Take some time to figure out what feels right – get some hands-on time with both the hardware and the operating system. Don't be swayed by splashy commercials, try not to focus on price, and absolutely do not lend any credence to the recommendations of store clerks.

You will need to make up your own mind, but here is some of our key advice to help steer you in the right direction.

Selecting A Smartphone Platform

Apple's iOS operating system powers iPhones and iPads. Google's Android powers most "smart" devices from other manufacturers – including Samsung, LG, Motorola, HTC, and a whole slew of low-cost bottom dwellers.

Amazon is also using Android to power the Kindle Fire, but Amazon has taken Android in such a divergent and incompatible direction from Google that it might as well be thought of as an entirely separate platform.

RIM (the maker of the old Blackberry) and Microsoft are still pushing smartphone platforms of their own, but they have had very little success winning over users and developers.

So at the moment it is a two-horse race, and for most people it comes down to choosing between iOS and Android.

Android is a great platform for advanced users who like to play around with configuring and customizing the software on their phones. You can do some legitimately awesome things tweaking an Android device to fit you like a glove, things that are often impossible on an iPhone.

But if you don't want to make maintaining and tweaking your phone a hobby, and if you want something that works reliably and that will be supported for a long time – unless you already know you have a preference for the Android ways of doing things, most average users tend to be better off going with Apple's iOS.

But spend some time getting hands-on with both platforms and see what is most intuitive to you. They both have strengths and weaknesses.

Since Androids have multiple manufacturers and devices, you have more hardware choices and cheaper price points.

Another big consideration is who you can go to for support on your device. If you have techno-savvy friends or family, sticking with their preferred platform ups the odds they'll be able to help out when you need some guidance.

A huge benefit for the iPhone is Apple's Genius Bars in their stores across the country. You can attend classes and get free support in person for any Apple product you own. There's no Android equivalent for device and OS support.

A note about contracts & subsidized devices: None of the carriers require a 2 year contract and most are in process of ditching them entirely. The standard now is EIPs (Equipment Installment Plans) or leasing. This separates out the device purchase cost from the monthly service plan.

In the past, the standard way to buy a phone was to sign a two-year contract – getting a huge discount up front, and the carrier makes back the cost (and more) over the course of your contract.

Device payments were technically bundled in with your monthly service cost.

If your phone is now out of contract, contact your carrier and see if you can reduce your plan costs. In most cases you can, but you have to ask.

Tips for Choosing an iPhone

Things are easy if you decide to go with iOS.

Because Apple's product line has been kept so simple, the choice is easy – buy whatever you can afford and what feels right in your hand.

Apple hardware tends to be really well built, and since the software is supported for so long, older phones tend to hold their value. If you aren't up for buying the latest and greatest, you can easily find a used phone that will still give years of good service.

Apple has a demonstrated track record of providing ongoing support for the hardware they sell – providing easy and free OS updates for years after launch, and free in-person tech support at Apple Stores.

As we write this (February 2016), we'd consider the iPhone 5 the oldest iPhone worth buying, and the iPhone 5S, 5C, 6, and 6S all absolutely awesome options.

If you like to be on the cutting edge, you can buy a new iPhone every year and trade in or sell your old iPhone to keep the incremental costs under control. Or Apple now offers their iPhone Upgrade program where instead of buying a phone, you pay a monthly fee and can upgrade annually.

Tips for Choosing an Android Phone

The Android operating system has begun to approach Apple's level of polish in the latest releases, but the quality of the hardware running Android is all over the map.

Some of the high-end Android flagships are every bit as solidly built as the iPhone – the Nexus 6P being a stellar example.

A lot of the lower-end Android devices, on the other hand, are so cheaply built that they are best avoided, no matter how low the selling price.

If you do go Android, please do not be tempted by these cheap generic phones or last year's now orphaned major-brand models. Invest in the latest Android flagship devices from Motorola, HTC, LG or Samsung and you will have the best possible experience going forward.

Make sure that whatever Android device you get ships with the current version of Google's Android operating system. Google uses candy-name code words in alphabetical order to indicate the version of Android – the recent ones have been named Ice Cream Sandwich (4.0), Jelly Bean (4.1 - 4.3), KitKat (4.4), Lollipop (5.x), and Marshmallow (6.0).

Beware buying any Android phone that ships with Jelly Bean or earlier – that these phones have not been upgraded to KitKat or better by now is a near-certain sign that they never will be, and they have been orphaned by their manufacturer.

Android phone makers customize the operating system with their own unique look and feel, with bonus features and bundled apps of sometimes very dubious value. If you want to experience Android as Google intended, look for phones branded "Nexus" or "Google Play Edition." These devices ship with a stock build of the Android OS and are able to receive OS updates directly from Google.

If you don't have a device that is being directly supported by Google, you are at the mercy of your cellular carrier and device manufacturer for OS and security updates. And a lot of them have a very lousy track record of following through on updates – there are millions of Android phones out there with known major security holes that will never, ever receive fixes.

Only the major brand flagships tend to get much ongoing support, especially a year after initial launch.

Locked & Unlocked Cellular Devices

SIM cards make it easy to move service between devices. If your phone is compatible with the underlying frequencies and cellular standard, by swapping SIM cards you can even change carriers. But only if your phone is unlocked.

Carriers very commonly lock new phones and hotspots so that they will only work on that carrier's own network. If you put in a SIM from a competitor, it will just not work.

Verizon is the only major carrier that currently has a policy of selling all phones and tablets unlocked – a very nice perk.

All of the carriers however are required to unlock your phone at your request once you have fulfilled any contract or device installment plan obligations. Prepaid phones tend to need to be in service for several months before they are released too.

Handle getting devices unlocked BEFORE you head out on an international trip or try to gift your old phone to a friend on a different carrier.

VoLTE / Simultaneous Voice & Data

LTE was designed from the ground up as a data network – and it actually does not have any built-in support for traditional voice phone calls. This used to mean that when a voice phone call was underway, you would actually drop off the LTE network!

This makes it frustrating to look up details or maps while on the phone making plans, and it can be especially frustrating if you are using your phone as a personal hotspot – chatting with a friend means that everyone else sharing your network is offline and twiddling their thumbs.

Cellular Data Gear

AT&T's and T-Mobile's 3G networks supports simultaneous voice and data – so on these carriers, this has never been a major issue. Network speeds drop down to 3G while a voice call is underway, but at least the internet connection stays up.

On Verizon and Sprint, on the other hand, the legacy voice network is completely separate from the data network, so while a voice call is underway all data connections cease.

Almost all newer Android smartphones haves gotten around this problem by building in an entirely separate second radio for voice calls so that the LTE data connection can remain online while a voice call is underway, but Apple never took this step with any iPhone models.

Why implement a temporary fix...when Voice over LTE (aka VoLTE) is emerging?

VoLTE is a technology that treats voice phone calls as just another data connection, meaning that the LTE network does not need to disconnect or switch into a special voice-only mode. VoLTE also opens the door to higher quality voice calls too, and to seamless switching between video and audio calls. It's now all just data.

All the major carriers except Sprint have begun to roll out VoLTE, and once VoLTE is everywhere, the whole idea of not having simultaneous voice and fast data will seem silly.

VoLTE also opens the door for cheaper LTE-only phones that can at last shed support for legacy voice, 2G, and 3G networks. This should reduce cost and complexity substantially, but LTE-only devices only make sense once LTE networks are 100% deployed.

Verizon's expansion into Alaska in 2015 was the first LTE only network that did not provide backward compatibility with older 3G-only devices, and in the future there will be more and more areas that may have LTE service but no legacy 2G or 3G support.

But for nomads – it is best to keep your options open and to make sure that you always retain the ability to fall back to older networks, which are often slow to get upgraded in some of the remote areas.

Wi-Fi Calling

Similar to VoLTE, Wi-Fi calling treats voice calls as data streams, but instead of sending them over the carrier's LTE network, the call is actually routed via local Wi-Fi networks.

Not surprisingly, T-Mobile and Sprint have been the most eager to support Wi-Fi calling, because it helps make up for their relative lack of cellular coverage. AT&T has also begun supporting Wi-Fi calling on iOS as of September 2015 – and more devices will be supported with time.

And even Verizon has started to embrace Wi-Fi calling, with very limited support towards the end of 2015, with more to come in 2016.

Wi-Fi calling opens up the door for using a cheaper cellular provider with more limited coverage for voice service, and using a data-only connection to help ensure that you can still get coverage in more places so that your voice calls and texts will still get through.

Wi-Fi calling requires both support from your carrier, and compatibility built into your phone.

SIM Cards & Migrating Service Between Devices

With all these device options, it's not uncommon for those utilizing cellular data to want to switch service between the devices in their arsenal.

Swapping SIM cards makes this possible.

All LTE-compatible devices have a tiny removable sliver of plastic called a SIM – short for Subscriber Identity Module. This little chip is what tells your carrier what device you want to use your plan with.

If you want to use your plan in another device, it's usually as simple as moving the SIM card over. However, not all plans are compatible with all devices. For instance, if mobile hotspot usage isn't a feature offered on the plan you're subscribed to, moving the SIM card from your smartphone to a MiFi may result in no internet service. And moving a SIM card associated with a data-only plan to a smartphone will not provide voice & texting service.

This card is often found lurking underneath the battery of many phones or MiFi devices, and can be carefully slid out with a fingernail. For devices that lack removable batteries, the SIM is usually in a tiny ejectable tray that can be removed by pushing a pin or paperclip carefully into a pinhole.

If you move the SIM to another compatible unlocked phone or device, you are essentially transferring your service (including your phone number) to that new device.

This can be extremely handy – you can take a SIM from a Verizon Mobile Hotspot and put it into an iPad, for example – getting data on a tablet that you can still share via the iPad's Personal Hotspot feature.

Cellular Data Gear

Or…you can keep an old beater phone around for when you are heading out into rough conditions – such as a backcountry hike or kayaking. Before you head out, just pop your SIM into it and you can still keep online and get phone calls, without putting your flagship expensive smartphone at risk.

It is quite common for those with grandfathered unlimited data plans from Verizon to move the SIM associated with the account to a MiFi to only utilize the data portion of the plan - making it easy to share the unlimited data.

Or…when you are traveling internationally, rather than paying expensive international roaming fees, you can instead get a SIM (and a local phone number) from a local cellular company, and then get online much, much cheaper.

SIMs have been getting smaller over the years to keep pace with ever-shrinking cell phones – the original SIM from the early 1990s was the size of a credit card.

But the ones you are most likely to see today are the Mini-SIM, Micro-SIM, and Nano-SIM. They are all electrically identical – so it is actually possible to cut down a Mini-SIM to put it into a device that has a Nano-SIM slot. And you can use a small plastic cradle to put a smaller SIM into Mini or Micro slot.

You can buy a SIM cutter tool to make the process easy.

SIM cards have always been a part of the GSM phone standard and have long been used by AT&T and T-Mobile and other GSM carriers. And SIM cards are a mandatory part of the LTE technology standard that all carriers are now using. But older 3G and earlier devices from Verizon and Sprint do not use SIM cards. If you have one of these earlier devices and want to move your service to another device, you have no choice but to call customer service and ask them to migrate service behind the scenes for you.

Cellular Signal Optimization

One of the problems with cellular internet is that the signal can be quite variable depending on many factors – your device, the tower location, weather, how many people are also using the tower, local terrain, nearby buildings, and even your own RV's construction.

You can however buy various external antennas and cellular boosters that can improve the situation.

Sometimes these are lifesavers and can make a finicky signal usable enough to get your work done.

And other times they might make no difference at all, or might even degrade your signal.

Here are the options for improving a weak cellular signal:

Placement: A phone in your pocket pressed up against a big dense bag of saltwater (you) is going to get a lot less signal than a tablet sitting on a desk, and both will be at a substantial disadvantage to a mobile hotspot placed in a window.

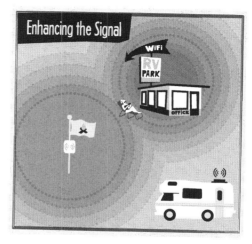

Sometimes just a slight relocation can make all the difference in the world – don't be afraid to experiment!

External Antennas: Very few phones and tablets have antenna ports anymore, but some mobile hotspots and USB modems do – and all commercial-grade equipment does.

If you have an antenna port, you can attach a more capable external antenna to that device for better reception. This is especially helpful if you want to keep your mobile hotspot set up in a fixed central location but allow for an antenna in a window or on the roof.

Some cellular devices support dual external antennas. Installing a matched pair of antennas preserves antenna diversity and allows for LTE's multiple-input, multiple-output (MIMO) capabilities.

Diversity provides for increased performance in noisy signal areas with a lot of signal reflections, and MIMO allows for multiple data streams to be combined for greatly enhanced speed.

To better understand what MIMO is and how it works – see the "Understanding MIMO" section of the "Wireless Signal Enhancing Tips" chapter.

Cellular Boosters: A cellular booster picks up a signal with an external antenna, amplifies it, and rebroadcasts it via an internal antenna to provide a stronger signal for all the nearby interior devices – whether they have an antenna port or not.

Boosters are frequency specific – you need a booster designed to work with the frequencies that your carrier uses.

We'll talk about cellular boosters later in this chapter in more detail.

Smart Boosters: A special type of smart booster actually functions more like a Wi-Fi repeater, only for cellular.

Smart boosters are carrier specific: Rather than just indiscriminately amplifying a raw signal, they know enough about the carrier's network to capture a signal and recreate it indoors. Because the booster is recreating the signal and not just amplifying it, it is able to ignore the noise – allowing for as much as a 100dB gain.

Smart boosters are at the moment only appropriate (and licensed for use in) fixed homes, but keep an eye out for possible future RV-friendly implementations.

Cel-Fi (www.cel-fi.com) is the most prominent company pushing smart booster technology, and they have told us a mobile version is on their longer term radar.

Cellular Signal: Bars & Dots vs. dBm

Everyone knows that more bars (or dots) is a good thing – but very few people realize that different phones and operating systems calculate how many signal bars to display very differently.

This means that comparing bars, unless you are on the same phone and same carrier, is actually a very poor way to compare signal quality between different devices.

The bars your phone is displaying sometimes do not even directly correspond to the actual signal strength: In addition to raw signal strength, the phone may be measuring network congestion and other variables to calculate how many bars to display.

In general, iPhones put more weight on network congestion when calculating how many bars (or dots, as of iOS 7) to display, and Android focuses more on raw signal strength.

This can make it harder to measure the impact a booster or antenna is having, because improving the signal strength might not register as more bars right away – especially if the phone is focusing on network congestion.

To objectively compare signal strength, you need to look for the dBm reading, which is based upon the actual power in milliwatts being received by an antenna.

This is a logarithmic scale – a 0dBm reading represent a single milliwatt received, and every change by 10 up or down represents a 10x change in received signal power.

The power levels picked up by a cellular antenna are fractions of a milliwatts, so the dBm readout will be a negative number.

–50dBm would be considered an awesome signal.

–60dBm is 10x weaker, but still great.

–70dBm is 100x weaker.

–80dBm is 1,000x weaker.

–90dBm is 10,000x weaker.

–100dBm is 100,000x weaker – and is when you are likely to start seeing a serious impact.

–110dBm is a million times weaker than –50dBm and is usually barely usable.

And by the time you see –120dBm, the phone has probably already given up and switched to "No Service."

Cellular Signal Optimization

Modern radios can work wonders with weak signals. In the past, −95dBm would have been considered weak, but a modern LTE radio can work with signals of −100dBm all the way down to −110dBm and often beyond.

Personally, we are blown away that a tiny, relatively affordable gadget that fits in your pocket can pick up signals so incredibly weak and do such amazing things with them.

The science and engineering involved here borders on magic.

See Real Signal Strength on an iPhone or Android

A hidden feature found in every iPhone running iOS 4.1 or newer is the ability to enable a special Field Test Mode that displays a real signal strength. This mode is enabled by going to the dial pad, and dialing: *3001#12345#*
When you quit this mode, the signal strength in dBm will be displayed briefly in the top left - replacing the signal dots.

On many Android phones, this reading can be found by going to Settings -> About Phone -> Status
If it's not there, you may need to dig around.

Cellular Boosters

A cellular booster provides several advantages to improve your connection:

1) An **external antenna** that you place on the top of your RV, or perhaps even in a window, is more capable and much better positioned than any built in antennas.

2) The signal is passed through an **amplifier** that makes faint distant signals louder and easier to pick up.

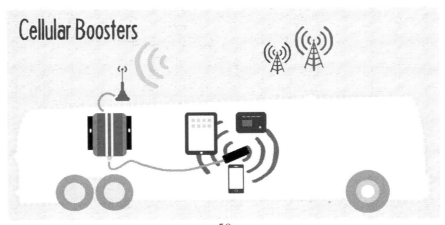

Cellular Boosters

58

3) An **interior antenna** rebroadcasts the amplified signal, allowing any cellular device within range of the interior antenna to receive an improved signal for both voice and cellular data.

4) When you transmit back to the cellular tower, this is all done in reverse – the **more powerful transmitter** inside the amplifier can substantially improve your upload speeds and connection reliability.

However, boosters are not miracle devices – they can't make signal out of nothing. There has to be some signal nearby for a booster to work with, and even then it may take some tweaking.

The maximum legal gain for a mobile cellular booster is 50dB, and for a home stationary booster the maximum gain is 70dB. RVs cross the line between being a home and being mobile. While stopped, RVs are stationary, and while in motion they are, well, mobile.

But generally speaking, the size of most RVs make the additional power of a 70db home booster difficult to work with, often causing problems with oscillation (when the exterior roof antenna picks up the signal transmitted from the interior antenna, instead of the tower.)

Unless special attention is paid to signal separation, we recommend mobile boosters for most RV applications.

When you are on the edge of a coverage area – or even dipped down into a canyon where cellular signal may not reach as strongly – this amount of gain can work wonders.

Some of the key manufacturers of cellular boosters include:

- weBoost (formerly known as Wilson Electronics) (www.weboost.com)

- SureCall (www.surecall.com)

- Maximum Signal (www.maximumsignal.net)

These devices can range from simple systems that just boost one mobile device at a time to complex systems that can provide RV-wide boosts for multiple devices simultaneously across multiple different carrier networks.

Keep Current Alert:

We keep this free resource center page updated as the cellular booster options change – start your shopping here for specific models & pricing:

Comparison: Mobile 4G Cellular Boosters
(www.rvmobileinternet.com/boosters/)

Booster Interference

If you are someplace where the signal isn't already weak or where there are many others nearby trying to get connected, a badly designed or improperly tuned booster is akin to a drunken loudmouth ruining the ambiance by shouting across the table in an intimate restaurant.

This potential for interference is real. And to the carriers who have spent billions of dollars buying up the rights to cellular spectrum, a consumer-grade booster stepping on their network's toes is not at all welcomed.

But at other times, carriers themselves are eager to recommend boosters if it helps them sell service to customers in fringe areas.

In early 2013, things at last came to a head, with the FCC working with cellular carriers and booster manufactures to jointly come up with new standards to prevent interference – yet to still allow boosters to be sold without needing special permission granted to each consumer.

The new rules went into effect in May 2014, and newly approved boosters finally hit the market in 2015 with built in technology to limit the potential for signal interference.

Most LTE-compatible boosters will comply with the new standards.

Cellular Booster Tips & Tricks

Here are some tips and tricks for getting the most out of a cellular booster:

- **Don't Boost When You Don't Need It:** Don't just leave your booster on all the time and assume that at every location you need it. At every stop, do some speed tests with the booster on and off to determine if you will benefit from it. If you have a good signal to begin with, a booster isn't going to make it better – and it could actually make things worse by confusing your device, crippling MIMO, or causing interference on the cellular network.

  This is less of an issue now with the new generation of FCC-approved boosters, but a pre-2014 booster always left on that may have been a help in remote areas could actually cause problems for the network in urban areas. The general rule of thumb is to turn off your booster when you don't need it to avoid potential interference.

- **Make Sure You Have a Proper Ground Plane:** Many mobile boosters are intended for use on an automobile with a metal roof, thus why the antenna is magnetically mounted. However, these antennas also depend on the metal roof to reflect the signal up into them. Unless you have a metal RV (such as an Airstream or bus conversion), more than likely your roof is a composite material or rubber that does not have the properties of a ground plane. You need to also install a piece of metal underneath the antenna. See the 'Antennas' chapter for more details.

- **Boosters Help Batteries Too:** Nothing drains a cellular device's battery faster than being in a fringe signal area. To stay connected, your gadgets need to operate their transmitters in full-power mode, and they are constantly searching for a better signal to lock onto. If a cellular booster is in the mix, your gadgets can transmit in low-power mode – vastly improving battery life.

If you know you are going to be in a fringe area and do not need to be constantly connected, put your phones into airplane mode. The battery will thank you!

(Ever wonder why, when driving through more remote areas, your phone's battery seems to drain quicker than normal? This is why!)

- **Know the Frequencies:** Even a booster sold as compatible with your carrier may not support all the frequencies your carrier uses. Know what your booster supports and what your carrier uses – and set your expectations accordingly. Sprint in particular uses LTE Band 41, which no booster can support.

(See the "Understanding Cellular Frequencies" chapter.)

- **Toggle Airplane Mode after Enabling a Booster:** Your mobile device occasionally may not notice that a new stronger signal has suddenly become available once you turn a booster on. To make sure that it does a full scan and finds it, after you turn on your booster it can help to reboot your device or toggle it into and then out of airplane mode.

- **It's Not Just About Bars & Dots** Especially on iOS devices, the signal bars displayed reflect network congestion as well as raw signal strength, and as such, enabling a cellular booster may not show up as more bars even if the signal is now stronger.

Instead of relying on bars, run a speed test to see what the actual difference in connection speed is.

- **Uploads Are Most Impacted:** Sometimes a booster may barely improve your download speed, but turning it on will make a huge impact on your upload speed. This is because the radios on cell towers are hugely powerful compared to the tiny radios integrated into mobile devices. If the tower is yelling at maximum volume, your device may be able to hear it just fine – but the tower may not be able to hear the tiny return whisper coming out of your phone.

A booster is akin to climbing on the roof and using a megaphone to yell louder – making it a lot easier for the tower to hear you. And this can vastly improve your uploads.

We've seen uploads go from a "barely can send a text" 50Kbps to a "let's video chat" 3Mbps just by toggling on a booster, even in places where the download speeds show barely any improvement at all.

- **LTE Isn't Always Better:** Even if your phone or hotspot defaults to LTE, try disabling LTE (if you can) to see the impact.

On Verizon, you will fall back to 3G – an almost always much worse experience. But AT&T's 4G HSPA+ network in our experience is often actually as fast or even *faster* than AT&T's LTE, particularly in fringe signal areas. You may occasionally experience the same improvement in speeds by going to a slower network on Sprint and T-Mobile too. Trial and error is the only way to know for sure.

We sometimes see weak LTE signals that even after boosting are not as usable as a strong nearby 4G signal.

- **Separation Matters:** Boosters thrive on distance between the inside and outside antennas to avoid the signal oscillation that can occur if the booster is picking up its own transmission.

The ideal distance between antennas for a 50dB amplifier (the most powerful approved by the FCC for mobile use) is actually 40 feet! That distance is impossible to achieve on most RVs, but it helps to do your best by separating antennas as much as is practical.

Because of the need for separation, a lot of more powerful home or small office cellular boosters will never be a good fit in an RV – they may just be too strong to avoid oscillation at all.

- **Aim Away:** To reduce the need for separation, some boosters come with directional antennas. Make sure that the outside antenna is pointed well away from the inside antenna to reduce your risks of oscillation and the booster shutting itself down.

You can do some experimentation to figure out which direction has the strongest signal to aim your outside antenna towards, or you can use some online tools to look up cellphone tower locations.

These lists of tower locations tend to be incomplete, but these apps and sites may help:

- Signal Finder (play.google.com/store/apps/details?id=com.akvelon.signaltracker) Android App

- OpenSignal.com

Cellular Signal Optimization

- **Never Boost Without an Antenna:** Boosters are not designed to operate without antennas connected – and on some, doing so will void the warranty and potentially fry the internal circuitry. If you are swapping antennas, *always* power down first!

- **Multiple Boosters Can Lead to Headaches:** It is sometimes handy to have different boosters on board, but be careful trying to use more than one at the same time. Do some experiments to make sure that they are not picking up on each other and actually degrading the signal instead of improving it.

- **Turn Off Gadgets You Don't Need to Boost:** If you leave all your tech with their cellular radios on, your booster can end up distracted trying to boost everything nearby at once.

We've observed that boosters often perform a lot better if other nearby devices that are not actively in use are put into airplane mode, so that the booster can focus on just one or two key devices. Turning off extraneous gadgets may be especially helpful if you are struggling to keep your booster from oscillating.

- **Beware Crossing Borders:** Some boosters are certified for both US and Canada use. If yours is not licensed for use in Canada or Mexico, you do not want to get caught using an unauthorized booster in a foreign country. Check with your manufacturer and purchase one certified for the countries you want to utilize them in.

Boosters licensed for use in the US will have an FCC certification number stamped on them, and boosters licensed for use in Canada with have an IC certification number. Some boosters may have both.

- **If You Get "The Knock" – Shut It Down:** It is not hard for carriers and the FCC to track down a malfunctioning booster that is causing interference with the local cellular network. We have spoken to a handful of RVers who have gotten a knock or letter telling them they need to shut down their booster until they can resolve the interference. You are required to comply with these requests, and the penalties can be severe if you ignore them.

If you get contacted, shut down your booster immediately. Contact your booster manufacturer or the vendor you purchased it through. In the past, we've heard of examples of free new units being sent out to replace a defective booster.

Theoretically, the new FCC-certified designs should be much less likely to cause any issues.

Can I Change Around Antennas?

The tiny magnetic mounted antennas that come with most cellular booster kits, believe it or not, are actually pretty well matched with the boosters and optimized for all of the current frequencies and carriers. Especially if they have a proper ground plane under them - they function well.

And by FCC regulations, if you use an antenna not approved by your booster manufacturer then you may be required to discontinue use if it's found to be interfering with a cellular tower's performance.

There is however physically nothing that prevents an end-user from changing around antennas at a later date, and the rules do allow for booster antenna upgrade kits to be sold.

In other words, if you want to use a different antenna with a booster, contact the manufacturer for advice and recommendations.

Unless you understand frequencies and antennas to know how to pick one out, we don't recommend purchasing just any random antenna. You can actually get negative results with some of them, including the popular dual band 'Trucker' style antennas which are best for voice & 3G signals – but can have negative gain on some LTE bands.

Refer to the "Antennas" chapter later in the book for more information.

A Signal Too Far

No matter how powerful a booster you have, sometimes you might be able to see a strong signal from a cell tower – but the tower still ignores you.

The timing of cellular signals is so precise that the speed of light intervenes, and if it takes too long for the tower to hear your device's reply, it moves on to listen to other signals.

You'll most likely encounter this in the mountains: If you have clear line of sight to a tower far away, you might seem to have a great signal – but nothing you send will ever go through.

Connections to GSM cell towers are actually impossible beyond 22 miles (35 kilometers) because of the timing required. LTE towers have limits too – but they are configurable and can be dialed up much farther in fringe areas.

But if you are beyond the range the network engineer planned for, the tower will always end up ignoring you – booster or not.

Old CDMA phones (Verizon & Sprint – pre-LTE) do not have a built-in distance limit, and 30–45 miles is possible in perfect conditions.

Expanding Interior Cellular Booster Coverage

To avoid oscillation, mobile-approved boosters come bundled with interior antennas which provide a very small circumference of boosted signal. In some cases, if your phone or hotspot is more than 3' away from the interior antenna, it will not benefit much from boosting at all.

If you absolutely want to have the most powerful booster possible with you on the road, offering boosted coverage all throughout your rig and even outside – it is going to be a challenge. If you are up for ignoring the FCC rules, juggling directional antennas, maximizing antenna separation, and dealing with the downsides – you could probably make a full-home residential booster work.

On the other hand, you could use this one simple trick to get inside coverage for phone calls and data, all throughout your rig.

Rather than trying to create a large boosted zone that covers your entire interior, set up a tech cabinet or desk area where you mount the booster's interior antenna to create a small boosted zone – and this is where you should set up your gadget-charging station.

Then use the Wi-Fi hotspot feature or a router to enable using your boosted connection everywhere in your rig. One gadget in the tech cabinet acting as the Wi-Fi hotspot will get all your other tech online, whether they are in the boosted zone or not. For voice, use a Bluetooth headset or speakerphone.

If you are the type of person who loves having a handset, you can even get a Bluetooth-enabled phone setup – and place a station on your nightstand and at your desk while your phone stays charging away in the tech cabinet, benefitting from maximum boost.

FCC Booster Registration Requirements

Rules from the FCC that went into effect on May 1, 2014 require that boosters now come with a scary mandated warning label:

BEFORE USE, you MUST REGISTER THIS DEVICE with your wireless provider and have your provider's consent. Most wireless providers consent to use of signal boosters. Some providers may not consent to the use of this device on their network. If you are unsure, contact your provider. You MUST operate this device with approved antennas and cables as specified by the manufacturer. Antennas MUST be installed at least 20 cm (8 inches) from any person. You MUST cease operating this device immediately if requested by the FCC or a licensed wireless service provider.

All the major carriers have already issued blanket consent for the use of the new generation of FCC-approved boosters on their network, so you don't need to ask any of the big four for permission. But that doesn't get you off the hook from registering.

Many of the currently shipping boosters don't come with any instructions on where to go and register – just a warning sticker saying you MUST.

Here are the links to each of the major carrier's registration pages:

- AT&T (www.attsignalbooster.com/)

- Verizon (www.verizonwireless.com/wcms/consumer/register-signal-booster.html)

- Sprint (www.sprint.com/legal/fcc_boosters.html)

- T-Mobile (www.signalboosterregistration.com)

The booster registration requirement has generated a lot of confusion. Here are some answers to the most frequently asked questions:

Why Register?

The primary purpose of the registration databases being built is to help with network troubleshooting issues. If a defective booster is wreaking havoc on the network, the registration info may help carriers track down and isolate the problem before it causes too much interference.

There really isn't a downside to registering, other than just a little bit of hassle.

What if I Don't Register?

You will not be fined or hauled off to jail. But you might be required to cease and desist if your booster is caught causing any network issues.

This general leniency only applies to consumer-level boosters. If you install a booster labeled "for industrial use" without having documented explicit permission from a carrier, you may be facing "penalties in excess of $100,000."

And if you ignore a request from the FCC or any licensed carrier to stop using a booster that is causing interference…well, then you are just asking for trouble.

Are Old Boosters Still Allowed?

Old (pre-2014) boosters have not been grandfathered in – but they have not been banned either.

Here is how Verizon explains it:

> *"Verizon also tentatively approves the use of consumer signal boosters that do not meet the new network protection standards. This approval is provided only for the boosters not causing interference and may be revoked if the particular booster or booster model is found to cause interference issues. To help avoid possible interference issues, however, Verizon recommends that customers who need signal boosters replace existing boosters as soon as possible with consumer signal boosters that meet the new network protection standards."*

How Do I Register as a Mobile Consumer?

All the registration forms request some subset of the following information: owner's name, operator's name (if different), contact phone number, booster make, model, and serial number, date of initial operation, and *installed location*.

Some of the forms ask whether the booster will be mobile or installed at a fixed location, but many of them seem to not have considered mobile users – especially mobile users without a fixed-location home base.

In those cases, we recommend using your mailing address.

What if I Have Multiple Devices on Multiple Networks?

The guidance from the FCC says that you should register with every carrier where you will regularly be connected. You need to register once per booster per carrier – it does not matter how many devices you are connecting.

What About Friends Who Use My Booster? Guests?

The FCC has ruled that it is perfectly fine for friends and visitors to take advantage of your booster without explicitly registering. But if you have housemates who are making regular use of your booster, they should register with their carrier too.

What about Cellular Antennas?

If you are just using an antenna without an amplifier/booster, there are no registration requirements.

If you use an antenna with a booster other than what the manufacture has shipped as part of the bundle or listed as an authorized upgrade option, you are giving up your blanket approval to operate granted by the FCC.

That would make it no different than if you were operating an older pre-2014 booster, and in these situations you are personally accountable for responsibly using your equipment. This is why booster manufacturers can't comment "on the record" about antenna combinations that might work well with their boosters.

And for similar reasons – cellular antenna manufacturers will be reluctant to talk officially about booster compatibility too.

Booster and third-party antenna combos can (and do) work, but if you are causing interference and the FCC comes knocking – you have to cease operation. It is essentially a one-strike warning rule – and if you keep causing interference you are the one liable, not the booster or antenna manufacturer

What About Registering Wi-Fi "Boosters"?

These new rules only apply to cellular boosters, not Wi-Fi repeating systems. So you do not need to do anything with products you might be using for your Wi-Fi signal enhancing – such as from WiFiRanger, Alfa, JefaTech, Ubiquiti, Wirie, etc.

A Lesson in Cellular Evolution

Cellular data connectivity has been evolving rapidly over the past decade, with an entire alphabet soup of technical standards and protocols behind the scenes pushing more bits faster with each new generation.

It is more than most mortals should need to worry about – so the carriers adopted "G" for "Generation" as a simple marketing shorthand. When you see terms like 2G, 3G, and 4G, that's all it means – 2nd generation, 3rd generation, etc.

Here's a handy little infographic we created that will hopefully illustrate the evolution of cellular data technology standards a little better:

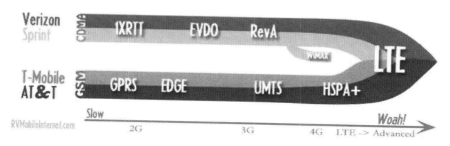

The Olden Days – 1G, 2G, 3G

It used to be that there were two competing and fundamentally very different wireless technologies: Sprint and Verizon used CDMA, and most of the rest of the world used GSM.

The third generation of CDMA technology was known as EVDO, and this enabled the first widely available usably fast wireless data speeds.

GSM networks were trailing at the time with 2G EDGE networks providing not much better than dial-up data speeds – fine for email, but painful for general surfing, and certainly not usable for video. But when the GSM networks evolved to 3G UMTS speeds, T-Mobile and AT&T leapfrogged ahead of CDMA's EVDO 3G speeds.

The future was bright for GSM networks AT&T and T-Mobile – with a clear technological evolutionary path mapped out from UMTS to even faster HSPA+ (3G+) to LTE (4G), with the network growing ever faster and able to handle increased user capacity.

CDMA networks, on the other hand, were at an evolutionary dead-end – there was no clear upgrade path for the carriers to anything beyond the slightly turbo-charged EVDO Rev-A. Forging ahead would require a totally new investment in core cell-towers and technology.

The 4G Revolution

Sprint bet big on a 4G technology called WiMAX and rushed to be the first to bring next generation 4G service to market. Embracing WiMAX as a successor to EVDO seemed like a reasonable bet years ago, but left Sprint headed down a technological dead-end.

Verizon predicted that the future was going to be LTE (aka Long-Term Evolution), and began aggressively building out the first and largest 4G LTE network in the United States.

Already being GSM based, AT&T should have had an easier time moving to LTE than Verizon, but AT&T has been perpetually playing catch-up.

To combat Verizon's lead and 4G LTE marketing push, in 2012 both AT&T and T-Mobile decided to start marketing their fastest HSPA+ areas as "4G" – generating some criticism and lots of confusion.

Meanwhile, seeing the LTE writing on the wall, Sprint stopped expanding its 4G WiMAX network and changed direction to focus on LTE as well. Sprint's WiMAX network was at last fully shut it down in November 2015 – orphaning all users of that once-promising technology.

Are we on the verge of grand-unified LTE nirvana?

Not quite.

Though all the carriers have been converging to use the same standardized LTE technology, all the carriers use different and incompatible radio frequencies – and you can only fit so many different radios and antennas into a device the size of a smartphone.

This is why there are often so many variants of particular phone models – the Sprint/Verizon versions still need to speak to legacy CDMA networks and talk to the Sprint and Verizon LTE frequency bands, while the AT&T or T-Mobile versions may leave the CDMA radio out entirely and support completely different LTE frequencies.

A few phones and devices include LTE radios compatible with all carriers – so if you are looking to hop easily between networks be sure to seek this capability out.

LTE-Advanced & Carrier Aggregation

Living up to its "Long Term Evolution" name, LTE networks were designed to continue to evolve with new capabilities and speeds while still remaining compatible with earlier LTE devices.

These evolved capabilities are known as LTE-Advanced, or LTE-A.

The most significant feature that LTE-Advanced has enabled so far is known as Carrier Aggregation, which lets an LTE-A radio combine multiple LTE channels from different discontiguous chunks of spectrum to create a higher bandwidth virtual channel that can support faster speeds.

Some of the newest LTE devices support two combined 20MHz channels for a peak theoretical cellular speed of 300Mbps, near-future devices will support three channels for 450Mbps, and eventually LTE-Advanced Pro devices will support more than ten data streams with at least 100MHz worth of total bandwidth – pushing cellular into gigabit speeds.

Think of it like a 100 lane highway through the sky!

The Fifth Generation (5G) Future

Beyond LTE-Advanced lies 5G.

The goal of 5G networks will be to enable ridiculously fast peak cellular data rates of over 10 Gbps, with network latency as low as 1ms.

This represents a 50x increase in network throughput and capacity compared to the fastest current 4G/LTE networks – and will truly represent a major generational shift when these technologies are at last deployed.

The actual technologies and frequency bands that will enable 5G have yet to be finalized by standards bodies – but 5G will likely take advantage of many disparate chunks of spectrum ranging from long range UHF frequencies up through short range extremely high 60 GHz frequencies.

Though true fifth generation cellular networks remain years away, research and development is well underway – as is the political haggling over what technologies will end up adopted as future industry standards.

Some experimental trial deployments of 5G technology by Verizon, AT&T, and Google have lately been making the news – and companies will be racing each other to be first to show off each major new advancement.

But even though 5G technologies will be increasingly in the news, the actual technical standards will not be finalized any sooner than 2019, with the first widespread commercial networks deployed no sooner than 2020.

The Need for Speed

Cellular companies love to brag about how fast their networks are – but once the connection is fast enough to stream HD video (around 5Mbps), until futuristic applications like augmented reality are common, do most mobile users really need anything faster?

Faster and more responsive surfing is nice – but with monthly data caps and potential overage charges, there are very real downsides to speeding down the information highway too quickly.

Often, we find ourselves wishing the network were actually slower so that we wouldn't accidentally burn through data so fast!

Why, then, are carriers so gung-ho about ever faster networks?

The key is capacity.

The faster the network is able to serve you whatever it is you've asked for, the faster it is able to get on to serving the next person. With only so much spectrum to go around and networks in many areas already oversaturated, more speed is almost a matter of survival.

Slower 2G and 3G data actually hog up more network airtime – costing carriers more to serve than sending the same data faster to LTE users. As the novelty of LTE wears off, expect carriers to make a big push to get users off the 3G networks and onto LTE as quickly as they can.

Verizon has already brought to market a few LTE-only devices which drop support for the old 3G and 2G networks entirely, and someday even today's LTE will be looking quant and old fashioned.

Understanding Cellular Frequencies

Every wireless broadcast travels along a radio wave with its own unique signature – its frequency.

The frequency of a radio wave is a measure of how many wave peaks there are per second – for example, a frequency of 700MHz means 700 million wave peaks per second are passing by.

Bandwidth is the range of frequencies covered by a channel. The wider the range of frequencies covered, the faster data can be sent over the air on a channel. For example, Verizon's primary LTE network uses the spectrum from 746MHz to 757MHz for cell tower transmission, providing a download channel with 11MHz bandwidth.

An Analogy: The overall radio spectrum can be thought of as vast tracts of real estate along one very long road, with frequencies being the addresses of various properties, channels being marked off by lot lines, and bandwidth representing how much street frontage each channel's lot covers.

The radio spectrum road ranges from the ultra-low frequencies where massive radios with miles-long antennas are used to communicate with

73

submerged submarines, up to the extremely high frequencies used for exotic scanners and sci-fi-style energy weapons.

Lower frequency radio waves require physically larger antennas — but are better able to travel long distances and penetrate through walls and obstructions. Higher frequencies on the other hand work well with small antennas, and are better suited for shorter range and higher speed line-of-sight transmissions.

The prime real estate in the middle of the radio spectrum is the UHF band (300MHz through 3GHz) — a sweet spot that allows for physically small radios that can still broadcast through walls and over relatively long distances.

This versatility makes for a crowded chunk of spectrum where digital television, Wi-Fi networking, aeronautical navigation, car alarms, walkie-talkies, satellite radio, cellular voice and data service, and even microwave ovens all carve out their own valuable pieces of real estate — squeezed in amongst dozens of other licensed and unlicensed users.

Unfortunately, there is only so much of this prime real estate to go around.

Cellular carriers have spent billions of dollars buying up spectrum (and each other) so that they have the capacity they need, and in some major urban areas the networks are already operating at near the saturation point.

> **An Analogy:** Think of cellular carriers like restaurants in business along the spectrum roadway. The more bandwidth a carrier owns, the more tables in the restaurant, and thus more customers they can seat at a time.

This UHF spectrum becoming overcrowded is in part why carriers are so afraid of offering unlimited data plans — there just isn't enough physical capacity to offer an unlimited buffet to insatiable consumers. There is simply more demand for meals at the cellular restaurants than there are tables to seat everyone!

A big part of the push towards faster networking technologies like LTE isn't just to offer customers faster speeds, it is to serve them faster so that a given chunk of spectrum can handle more simultaneous users.

Just like a restaurant — the faster you can turn around the tables, the more customers a business can handle in a day.

As cellular data has grown more and more valuable and important, previously occupied areas of spectrum are being cleared to open up more room for cellular usage.

Examples:

- The 700MHz frequencies used by Verizon and AT&T's LTE networks used to be analog television channels 52–69, and this spectrum was only made available for cellular usage in 2009 as part of the transition to digital television.

- The 1700MHz/2100MHz AWS frequencies that many carriers now offer LTE service over were freed up by shutting down a wireless cable TV service that never really caught on.

- In early 2016 the government is preparing to facilitate a big incentive auction in the 600MHz bands, allowing the current licensees of television channels 38–51 to voluntarily sell off their spectrum holdings so that they can vacate and make even more space for cellular data expansion.

The bulldozers are getting ready to move in, but it will still take a few years before cellular networks utilizing this newly cleared bandwidth are fully built out and compatible devices are released.

Until then, the cellular restaurants remain crowded with customers willing to pay by the meal, so truly unlimited data buffets will remain scarce.

Cellular Frequency Bands

There used to be a time when there were only two neighboring chunks of 850MHz spectrum dedicated to cellular usage, allowing for just two cellular carriers in any given region.

But things have gotten a lot more complicated as the move to 2G and digital opened up more potential channels, and Sprint pioneered this expansion by launching its nationwide network in 1995 on the 1900MHz PCS bands.

Today there are nearly a dozen defined LTE bands in active use in the United States, with more being added. Every carrier owns multiple blocks of spectrum within various combinations of these bands – but not all phones and data devices actually have radios capable of tuning in to every frequency and communications standard.

A smart shopper wanting access to a carrier's entire network should make sure that his or her cellular devices support all the current and even the planned future frequencies used by their selected carrier.

Remember – lower frequencies tend to provide more range, and higher frequencies tend to provide more speed. It is important to have a carrier and devices that support both extremes.

Understanding Cellular Frequencies

Here are the current cellular frequencies in use in the United States:

Frequency	Common Name	Band	Who Is Using It?
600MHz	UHF TV Channels 38-51	tbd	Auction happening in 2016.
700MHz (Lower)	700MHz Auction Block A/B/C	LTE Band 12	T-Mobile "Extended Range LTE"
700MHz (Lower)	700MHz Auction Block B/C (Subset of band 12)	LTE Band 17	AT&T LTE
700MHz (Lower)	700MHz Auction Block D/E (Download Only)	LTE Band 29	AT&T LTE – Coming in 2016.
700MHz (Upper)	700MHz Auction Block C (Upper)	LTE Band 13	Verizon LTE
850MHz	Extended CLR (Cellular)	LTE Band 26	Sprint "Spark" LTE
850MHz	CLR (Cellular – the original!)	LTE Band 5	Verizon 3G, AT&T 3G, AT&T 4G
1700MHz/ 2100MHz	AWS (Advanced Wireless Service)	LTE Band 4	Verizon XLTE, AT&T LTE, T-Mobile 4G, T-Mobile LTE
1700MHz/ 2100MHz	AWS-3 (Advanced Wireless Service – 2014 Auction Expansion)	LTE Band 66	Auction winners: AT&T, Verizon, T-Mobile, Dish
1900MHz	PCS (Personal Communications Service)	LTE Band 2	Verizon 3G, LTE; AT&T 3G,4 G, LTE; T-Mobile 2G,4G,LTE
1900MHz	Extended PCS	LTE Band 25	Sprint 3G, LTE
2.3GHz	WCS (Wireless Communication Service)	LTE Band 30	AT&T LTE (New in 2015)
2.5GHz	BRS (Broadband Radio Service)	LTE Band 41	Sprint "LTE Plus"
5GHz	LTE-U (Unlicensed) / LTE-LAA (License Assisted Access)	LTE Bands 252, 255	Experimental: Verizon, T-Mobile

That is a lot of frequencies – and it is fiendishly complex to build a phone, booster, or any other wireless device that supports many of these at once.

This is one of the reasons why many phones come in carrier-specific models and why roaming across carriers and internationally is often limited to the lowest common denominator frequency bands.

With all these incompatible frequency bands, things inevitably get messy – for example, the iPhone did not support T-Mobile's AWS 4G bands until the iPhone 5 came along. You could buy an iPhone 4S on T-Mobile, but you would only ever get 2G data speeds in most places – no matter what the coverage maps said.

And up until late 2015, very few devices supported LTE Band 12 – the spectrum T-Mobile has been using for "Extended Range LTE" to massively expand its coverage map. Because of this – an iPhone 6S will get T-Mobile LTE coverage in vastly more places than a year older iPhone 6 will.

These are just a few examples of the growing pains going on out there.

It is almost miraculous that so many mobile devices work so well considering all the chaos and complexity behind the scenes.

Carpool Lanes in The Sky

It helps to think of LTE spectrum like lanes on a highway.

The more lanes a carrier can offer, the more simultaneous users can be supported, and the faster each user can go.

Older devices without support for the latest LTE network bands are limited to the most congested lanes on the road, while newer devices can often zip past in the carpool lane leaving older devices stuck in traffic.

This is one of the reasons why it is important to keep your connectivity arsenal up to date.

Carrier Frequency Reference Guide

Verizon's Cellular Network

- **700MHz Upper Block (LTE Band 13)** – Primary LTE network.
- **850MHz Cellular (LTE Band 5)** – Currently supports voice calls, 2G 1xRTT, 3G EVDO. Eventual upgrade to LTE.
- **1700MHz/2100MHz AWS (LTE Band 4)** – XLTE
- **1900MHz PCS (LTE Band 2)** – Currently supports voice calls, 2G 1xRTT, 3G EVDO. Migrating to LTE rapidly.

Verizon's Future Plans

Verizon has been rolling out what it is calling XLTE on the AWS frequency bands across its network to increase speed and capacity, and now most new Verizon devices ship with XLTE support.

In many areas Verizon's 700MHz LTE network is overloaded and slow while the XLTE bands are still uncrowded. If you do not want to get left behind, make sure that all your Verizon tech is XLTE compatible – and if you don't have XLTE support, an upgrade is very worthwhile.

Verizon has also begun to convert (aka refarm) its 1900MHz PCS network from 3G to LTE – and will continue to do so across the nation. Verizon has a rough target date of 2021 for fully shutting down its old 2G and 3G networks.

In areas that have already been converted, half of Verizon's 3G capacity has been turned off to make space for LTE – so if you are still using 3G you may notice speeds and coverage getting increasingly worse.

To increase speeds – in 2015 Verizon began deploying LTE-Advanced carrier aggregation that allows compatible devices to combine two channels for faster speeds, and this will become increasingly important.

Looking further ahead – Verizon invested heavily in the AWS-3 spectrum auction (LTE Band 66), and has begun experimenting with 5GHz LTE-U. No current Verizon devices yet support either of these LTE bands, but by the end of 2016 Verizon's future rollout plans may be revealed.

Verizon has also said that it intends to lead the way in deploying 5G network trials in the future.

For maximum future compatibility – make sure that your Verizon devices support LTE Bands 2, 4, 5, and 13 – and that they have support for Verizon's legacy CDMA 3G network too. Carrier aggregation will become increasingly important, but is only found on the latest flagship devices for now.

AT&T's Cellular Network

- **700MHz Lower Block (LTE Band 17)** – Primary LTE.
- **850MHz Cellular (LTE Band 5)** – Supports 2G GSM/GPRS/ EDGE, 3G UMTS, and 4G HSPA+. LTE now rolling out here.
- **1700MHz/2100MHz AWS (LTE Band 4)** – LTE deployed.
- **1900MHz PCS (LTE Band 2)** – Supports 2G GSM/GPRS/ EDGE, 3G UMTS, 4G HSPA+, and 4G/LTE in select markets.
- **2300MHz WCS (LTE Band 30)** – New LTE expansion in 2015.

AT&T's Future Plans

AT&T has announced plans to fully shut down its 2G GSM network by January 1st 2017, meaning that older devices from before the 3G era will no longer work after this date. This freed-up network bandwidth will certainly go towards extending LTE capacity.

AT&T has also started to refarm its 1900MHz PCS network to support LTE, with several major urban areas converted as 3G service is dialed back.

AT&T has also been buying up 1700/2100MHz AWS spectrum, which it has begun using in several areas, including using LTE-Advanced carrier aggregation which combines channels from different chunks of spectrum to make a higher bandwidth virtual channel that can support faster speeds.

Also taking advantage of carrier aggregation will be AT&T's new download-only frequencies in the 700MHz LTE Band 29. These channels need to be combined via carrier aggregation with bidirectional upload/ download bands to function, but it opens up a lot of extra capacity.

AT&T hasn't started using Band 29 spectrum yet, but has started to ship a few devices with support built in – meaning that these frequencies are likely to go into use sometime in 2016.

In 2015 AT&T began to expand LTE service into the 2300MHz WCS channels (LTE Band 30), and began rolling out service in limited areas.

AT&T also invested heavily in 2014's AWS-3 (LTE Band 66) auction, buying up more spectrum than any of the other bidders. It will take several years before the AWS-3 bands are ready for use however.

For maximum future compatibility – make sure that your AT&T devices support LTE Bands 2, 4, 5, and 17 – and keep an eye out for devices that support LTE Bands 29 and 30. Be sure to check for AT&T's carrier aggregation features too for maximum future proofing.

T-Mobile's Cellular Network

- **700MHz Lower Block (LTE Band 12)** – "Extended Range LTE" – widely deployed in 2015.

- **1700MHz/2100MHz AWS (LTE Band 4)** – Primary home of T-Mobile's LTE service and "Wideband LTE". Older HSPA+ 4G service being relocated to 1900MHz.

- **1900MHz PCS (LTE Band 2)** – Supports 2G GSM/GPRS/EDGE, and now 4G HSPA+ service. Being transitioned to LTE, with 2G being phased out.

T-Mobile's Future Plans

T-Mobile used to suffer from a notorious lack of coverage away from core metro areas, made substantially worse because T-Mobile relied on the relative high-frequency AWS bands, with little low-frequency spectrum better suited for long-range reception or building penetration.

To improve the situation, T-Mobile has acquired a lot of 700MHz LTE Band 12 spectrum, and in late 2014 began rolling out 700MHz service to better compete with Verizon and AT&T.

Thanks to Band 12, over the course of 2015 T-Mobile more than doubled its nationwide coverage area. There is a catch – very few mobile devices support LTE Band 12 yet.

In the places it has strong coverage, T-Mobile has an extremely fast network. T-Mobile labels places where it has deployed extra large bandwidth LTE channels as "Wideband LTE" markets. T-Mobile has also begun to support LTE-Advanced Carrier Aggregation to combine multiple channels together for even faster speeds.

T-Mobile has a 7-year roaming agreement in place to allow for service on AT&T's 850MHz 3G network. This fills a lot of the coverage gap for voice calls, also helped by T-Mobile's lead in deploying Wi-Fi Calling (VoWiFi) capability.

Looking ahead – T-Mobile invested in the AWS-3 (LTE Band 66) spectrum auction, and is publicly gearing up to make a big play in 2016's 600MHz spectrum auction. T-Mobile is also testing 5GHz LTE-U service, which expands LTE service onto the unlicensed spectrum normally used for 5GHz Wi-Fi.

For maximum future compatibility – make sure that your T-Mobile devices support VoWiFi, and LTE Bands 2, 4, and 12 – and that they have support for 850MHz UMTS so that they can continue to roam onto AT&T. For the future LTE Band 12 and carrier aggregations will be increasingly important.

Sprint's Cellular Network

- **800MHz Cellular (LTE Band 26)** – Voice, 2G 1xRTT, and LTE.

- **1900MHz PCS (LTE Band 25)** – Currently supports voice, 2G 1xRTT, 3G EVDO, and is Sprint's primary LTE band.

- **2.5GHz BRS (LTE Band 41)** – "LTE Plus"

Sprint's Future Plans

Sprint is pursuing deploying three LTE bands that, when combined, offer the potential for the best of all worlds – high-frequency speeds and low-frequency range.

Sprint originally called this tri-band LTE network Sprint Spark, but now that Sprint has started to release devices with carrier aggregation capabilities to use multiple bands simultaneously – it has taken to calling areas where this technology is deployed "LTE Plus" markets.

Sprint owns a huge wealth of undeveloped spectrum, but it has been struggling to expand to take advantage of it. At the time of this book's publications, Sprint had rolled out LTE Plus capability to only the core areas of 103 cities.

In some core urban areas, Sprint's network can deliver truly stunning speeds.

But in much of the rest of the country, Sprint has slipped into an increasingly distant fourth place when it comes to coverage, speed, and overall reliability.

For maximum future compatibility – make sure that your Sprint devices are "LTE Plus" compatible with support for carrier aggregations and LTE Bands 25, 26, and 41. Shop smart – as of the end of 2015, only a handful of devices are fully LTE Plus capable.

"Frequency" Asked Questions

Are some frequencies better than others?

Frequencies have trade-offs – lower frequencies travel farther and can more easily reach into the interiors of buildings.

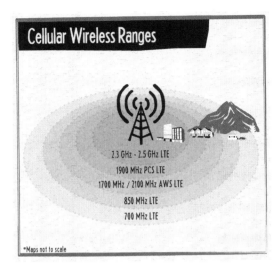

Higher frequencies tend to have more higher-bandwidth channels available, and thus offer faster speeds and increased capacity.

Carriers lacking in lower frequency spectrum have a much harder time offering coverage in remote rural areas since it takes a lot more cellular towers to cover the same amount of territory.

All the carriers are aspiring to build out multi-tiered balanced networks, with a mix of lower frequency spectrum for range and better building penetration, and higher frequency spectrum for speed and urban capacity.

But we remain in early days, and the race to bring out these enhanced networks is still just beginning.

Why does the AWS band have two frequencies?

The AWS band is split into two chunks – with downstream data from the cell tower sent at 2100MHz, and upstream data flowing back to the tower broadcast at 1700MHz.

Other cellular bands also separate uplink and downlink frequencies as well, but only in the AWS band are they so substantially far apart as to merit having both frequencies written out explicitly.

Why do some carriers brag about having more LTE bandwidth?

LTE signals allows for 1.4MHz, 3MHz, 5MHz, 10MHz, 15MHz, and 20MHz bandwidth blocks to be devoted to data transmission – usually with a paired upstream and downstream connection described as 5x5 or 15x15 or 20x20.

Understanding Cellular Frequencies

With more bandwidth devoted to LTE, the network can operate faster and more users can be served simultaneously.

Think of bandwidth in this context as having more lanes on a highway.

As a rule of thumb – an LTE network can serve 200 active full-speed users on each cell for every 5MHz of bandwidth dedicated. If a network becomes overloaded, however, performance suffers for everyone.

Theoretical Peak LTE Data Rates

Bandwidth	Download Speed	Upload Speed
5MHz	37 Mbps	18 Mbps
10MHz	73 Mbps	36 Mbps
20MHz	150 Mbps	75 Mbps

Carriers bragging about 10MHz, 15MHz, or 20MHz channels are showing off how much headroom they have – and how much raw potential speed they can offer.

Many smaller carriers with legacy networks have only barely unleashed LTE, devoting just 1.4MHz or 3MHz of spectrum to LTE until they have transitioned enough users so that they can afford to take spectrum away from older standards.

Verizon, on the other hand, has been able to dedicate at least 10MHz to LTE in every market in the continental United States. In contrast, AT&T only controls enough spectrum to currently offer 5MHz LTE universally, though in many places it has been able to expand to 10MHz and beyond.

What is Carrier Aggregation? How does it impact speed?

The LTE standard is limited to 20MHz bandwidth per channel, but the LTE-Advanced evolution of the standard allows for up to 100MHz bandwidth by combining smaller channels from different noncontiguous chunks of spectrum using a technology called Carrier Aggregation (CA).

Two 20MHz channels combined allow for a theoretical maximum speed of 300 Mbps, and in 2016 we may begin to see devices and networks which support three channel carrier aggregation for a maximum theoretical speed of 450 Mbps.

How do I know a device I am buying fully supports my carrier?

Check the frequencies and standards listed on the specifications for the device to make sure that it supports all the current networks deployed by your carrier.

Understanding Cellular Frequencies

Checking the LTE bands makes it easy, and be sure to check out the carrier reference guide in this chapter to be sure all your bases are covered.

What happens if I buy a device that does not fully support my carrier's network?

Here's an example that shows why it is important to look closely at the specs – particularly if you are buying an older device:

T-Mobile's LTE network is primarily on LTE Band 4, its 3G/4G network is HSPA+ 1700/2100MHz, and its 2G network is 1900MHz GSM EDGE. Some T-Mobile areas have had the 1900MHz EDGE switched to 4G 1900MHz HSPA+, which is compatible with a wider range of devices. But many areas have not.

If you look carefully at the specs, you will notice that some popular devices (such as the non-Retina iPad Mini) are NOT compatible with T-Mobile's HSPA+ 1700/2100MHz 4G network.

This means that in some T-Mobile 4G areas, non-Retina iPad Mini users will end up dropping back to super-slow 2G EDGE speeds.

The Retina iPad Mini, on the other hand, can handle every frequency that T-Mobile has in the US other than LTE Band 12, so Retina Mini users will often get fast 4G where Classic Mini users get slow 2G.

Only by knowing your frequencies will you be able to make a smart choice and avoid getting stuck unintentionally in the slow lane.

What if I want to roam internationally?

The GSM standard is by far the most deployed internationally – and quad-band World Phones are capable of roaming with at least 2G data speeds just about everywhere. Look for GSM 850MHz/900MHz/1800MHz/1900MHz support on your phone's or tablet's spec sheet. This used to be rare but is now very common.

LTE roaming is trickier – there are 44 different LTE bands in use around the world, and finding hardware that works on both your home network and where you want to visit may be tricky.

If your plan is to find a local SIM, the latest flagship phones are the ones most likely to offer support for the most global LTE bands. The iPhone 6S in particular is well suited to international LTE roaming with 23 different LTE bands supported.

What about cellular booster frequencies?

Most older boosters are dual-band boosting – supporting just the 850MHz cellular band and the 1900MHz PCS band. This typically helps with voice and basic data for all the major carriers, but offers little help for the majority of LTE signals.

A few newer boosters are tri-band – coming in versions made specifically for Verizon, AT&T, or T-Mobile. In addition to the dual bands, the Verizon tri-band boosters add support for the upper 700MHz band, the AT&T versions add support for the lower 700MHz band, and T-Mobile models add support for AWS.

Considering that Verizon and AT&T are also both rolling out service in the AWS band, getting a booster that lacks this support will prove to be short-sighted.

There are, however, five-band boosters that offer support for all of the above in one device. If you have devices from multiple carriers, a five-band booster is especially wise.

What on Earth is refarming?

The curious word "refarming" refers to the practice of shuffling around existing network frequency allocations to make room for new services and more efficient modern technologies.

In the process, older 2G and 3G technologies end up with slower speeds and less coverage allocated to them, or service may be eliminated entirely.

If you happen to still have an old device that is going to be made obsolete and unusable by network refarming, carriers have been known to offer free hardware upgrades to keep from losing a customer.

Wi-Fi Hotspots

Often the fastest, cheapest, and easiest way to get online is to use public Wi-Fi networks, and these are pretty easy to find.

Many libraries, coffee shops, RV parks, breweries (yay!), motels, municipal parks, and even fast food restaurants now offer free Wi-Fi. There are also plenty of paid Wi-Fi networks to be found too.

Though Wi-Fi has the potential to be blazingly fast, some shared networks can be horribly overloaded. A public Wi-Fi hotspot is highly dependent upon its upstream source of internet (cable, DSL, satellite, etc.) and on how many people are sharing that connection.

WiFi Hotspot

Quick Glance

Pros

Widely available

Easily accessible

Frequently free

Cons

Variable quality

Frequently unreliable

Security concerns

In some cases, the upstream connection may actually be little better than old dial-up modems. In some remote places, the upstream connection may actually BE a dial-up modem!

Unfortunately – in many situations, even though you may be able to get online via Wi-Fi – it may not be worth bothering with.

The other major limitation of Wi-Fi is range. Sometimes we enjoy working in a cafe or brewery, but usually we prefer to be at our home office or computing outside under the shade of a tree. Most Wi-Fi hotspots fall off to unusably slow connections just a hundred feet away from the base station, and in some RV parks only the nearest spots to the front office can reliably connect via Wi-Fi.

But with an external Wi-Fi antenna and/or Wi-Fi repeater system (see the next chapter), you can often manage to connect to a base station substantially farther away than your unaided laptop or tablet alone ever could.

Realities of Campground Wi-Fi

Although you would think that a campground that advertises "Free Wi-Fi!" as prominently as it does 50A power hook-ups would actually have worthwhile Wi-Fi, we have sadly discovered that this is often not the case. And once you understand all that is involved in providing a fast free Wi-Fi network, you may realize how unrealistic it is to expect that.

Maintaining a public Wi-Fi system that can serve hundreds of bandwidth-hungry travelers, especially if spread out over several acres (such as at an RV park), is very expensive to set up and maintain. Few RV park managers have the expertise to upkeep such a network, especially without having an IT expert on call.

To do it right requires a substantial investment in routers, repeaters, and equipment – not to mention, needing a pretty hefty internet backbone to tap into.

Generally, if the Wi-Fi is managed decently enough, it is common for RV park Wi-Fi to be good enough for checking email and doing some basic surfing – generally all that most RV campground patrons are assumed to really need.

But if all you are going to do is check email and some light surfing – it is hardly worth trying Wi-Fi when a cellular plan could do it just as well.

Wi-Fi Hotspots

On rare occasions, we've been at campgrounds with great Wi-Fi – fast and without data caps! It has occasionally been so good we could suspend our cellular account for the month.

But it is much more common for the signal or configuration at campground Wi-Fi installations to be so iffy that it's actually pretty frustrating.

All too often, campgrounds suffer seriously overloaded connections – particularly in the evening when a lot of people try to get online at once. A connection that might be decent during the day, while everyone is off exploring, might feel worse than dial-up during prime time. It can take just one or two people trying to stream a movie, video chatting with the grandkids, or downloading a huge file to bring the entire network to a grinding halt.

Some campgrounds have outsourced the chore of providing Wi-Fi to a provider, like Tengo Internet (www.tengointernet.com), who manages the bandwidth and network for them. Sometimes they even charge extra for it, cap how much data you can use, or limit how many devices you can connect at once.

The theory is that these limits help fairly spread out the available capacity.

You would think that paid and professionally managed Wi-Fi connections would end up being faster and more reliable than open and free, but in our experience this has not proven to be the case. All too often, the paid networks have proven to be a waste of effort, not even worth bothering to jump through the hoops to get online.

In the end, we look at campground-provided Wi-Fi to be a bonus if it's usable – but we've learned not to rely on it for anything critical and bandwidth intensive.

If campground-provided Wi-Fi is important to you, read online reviews at places like www.campendium.com, www.rvparkreview.com or www.rvparking.com to help in selecting where you head – a lot of folks comment on the reliability of the Wi-Fi in their park reviews.

And if you find the Wi-Fi not cooperating, try asking the front desk to reboot the router – that's a simple thing that can sometimes make huge improvement.

Free Public Hotspots

A lot of businesses and public resources provide free Wi-Fi. For example – buy a cup of coffee, and you can spend an afternoon using a cafe's hotspot for your laptop. Walmart's in that past year have been seen offering more free Wi-Fi hotspots, which goes along nicely with a free place to park overnight while in route (at locations that allow it).

Heck, even many McDonald's offer free Wi-Fi to go with your super-sized fries. "Would you like bandwidth with that?"

If you have the flexibility and/or desire to take your laptop with you, these can be great supplements to your internet arsenal. Some folks actually rely on this method as their primary internet. And if you can get your RV within range of these hotspots, you might even be able to use them from the comfort of your recliner.

Places known for their free hotspots include libraries, laundromats, McDonald's, coffee shops, Panera Bread, Lowe's (yes, the hardware store), Walmart, some rest stops, motels (higher end hotels tend to have paid internet), breweries, restaurants, and so many more.

There are several apps and websites out there that you can use to track public hotspots down – but we usually find it pretty easy to stumble into these places when we need them.

You might have to ask for the password, but many places are happy to share their network with their customers.

Some days we find a restaurant with free Wi-Fi to eat a late lunch at, and then spend the afternoon while the tables aren't in high demand working away. Some places are even happy to seat you near a power outlet!

Some other places limit how long they'll let you stay on their connection, as they do need their tables for customers just arriving. If asked to move on, be courteous and comply.

These businesses provide these services as a courtesy to their customers – please return the favor by being their customer. Order food and beverages, and tip your server well for taking up a seat in their area for an extended amount of time.

And always remember, not all Wi-Fi is created equal and your speeds may vary from location to location. Before buying a meal or coffee, we run a speed test from our devices to make sure the bandwidth offered is usable enough for our needs.

Paid Wi-Fi Networks

There are some widely deployed Wi-Fi networks out there available to paid customers.

For example – Florida cable company Brighthouse is installing a network of Wi-Fi hotspots across Central Florida for their customers. This is part of a

Wi-Fi Hotspots

nationwide Cable WiFi (www.cablewifi.com) initiative involving other regional cable companies (including Time Warner Cable, Cox, Optimum, and Comcast's Xfinity), all of which allow customers to roam freely between connected Cable WiFi hotspots.

With over 300,000 Wi-Fi hotspots and growing, if you are in an area served by one of these companies you might be surprised to find that you can get access in some very unexpected places – if you are an authorized user.

Check if your stationary family members and friends are customers of one of the participating cable companies – you might be able to get their permission to use their login to access the network.

Comcast's Xfinity WiFi (wifi.comcast.com) is taking things even further – and now has millions of "xfinitywifi" hotspots. Comcast has accomplished this by turning business and home customers' cable modems into public Wi-Fi hotspots – in one fell swoop offering fast Wi-Fi over entire neighborhoods.

Non-Xfinity customers can get two 60-minute complementary Wi-Fi sessions a month, or you can buy hourly, daily or weekly passes.

If you see "xfinitywifi" or "CableWiFi" as an available hotspot, you can try this out as an option. If you are in a Comcast area, the Xfinity iOS and Android apps will help you find areas that are covered.

There's also services like Boingo (www.boingo.com), which has over one million hotspots around the world – including many Tengo location found within campgrounds. For a monthly fee generally less than subscribing to Tengo, you can get unlimited access at all of their locations.

Borrowing Bandwidth From Friends

Our favorite way to access Wi-Fi is by borrowing a cup of bandwidth from friends and family as we travel.

We find most folks with fast home connections are more than happy to share their unlimited high-speed bandwidth when we need to do things like OS updates, download shows from iTunes or Amazon Prime, get the latest development tools, or do a massive backup to DropBox.

We keep small meaningful gifts on board to thank our gracious hosts.

Wi-Fi Congestion Issues

Most Wi-Fi devices operate on an unlicensed 2.4GHz frequency band that provides for only THREE fully distinct radio channels (en.wikipedia.org/wiki/List_of_WLAN_channels)

> TIP: If you are manually configuring a Wi-Fi device, channels 1, 6, and 11 are the ones that do not overlap and interfere with each other.

Bluetooth devices and most cordless phones (remember those?) also operate in this same unlicensed 2.4GHz frequency band – further adding to the congestion. In an urban area or even a crowded campground, those three channels can get awfully overloaded with signals, all interfering with each other.

But it is even worse than that: Microwave ovens also emit radiation in the same 2.4GHz band, meaning that if your neighbor is making popcorn, it can potentially grind Wi-Fi speeds to a halt for everyone nearby if the microwave isn't well shielded.

Despite all the interference, it is amazing how well 2.4GHz Wi-Fi works.

But with so much traffic and more and more Wi-Fi devices in use every day, using Wi-Fi on 2.4GHz is like trying to have a conversation in a loud bar while a heavy metal band is playing and a jackhammer is outside.

5Ghz Wi-Fi – An Uncrowded Expressway

There are also Wi-Fi channels located in the uncongested 5GHz frequency band, where 23+ nonoverlapping channels are sitting usually vacant. It is the express lane compared to the 2.4GHz gridlock – a quiet library compared to a raucous bar. But to take advantage of 5GHz requires different antennas and equipment specifically designed to broadcast on these channels, and even now many new Wi-Fi devices skimp on 5GHz support.

It has been a classic chicken-and-egg problem – manufactures need to include 2.4GHz support for backwards compatibility for communicating with existing devices, and since there are so few 5GHz devices, manufacturers are tempted to keep costs lower on new devices, leaving 5GHz support out.

Wi-Fi Standards & Ranges

Wi-Fi is defined by the IEEE 802.11 (en.wikipedia.org/wiki/IEEE_802.11) set of standards – and the specifications for any Wi-Fi-compatible device should indicate which variants (indicated by letters appended to 802.11, such as 802.11b/g/n) of the standard are supported.

When two Wi-Fi devices try to connect with each other, theoretically they should negotiate the fastest and most recent connection standard that they are both compatible with.

Here are the Wi-Fi standards that you might run across:

- **802.11b** – The original 2.4GHz Wi-Fi standard brought wireless networking to laptops everywhere starting in 1999. It has a maximum speed of 11Mbps, and is mostly obsolete now.

- **802.11a** – The original 5GHz Wi-Fi standard also came out in 1999, but it was never widely deployed in consumer gear.

- **802.11g** – The next 2.4GHz Wi-Fi standard upped network speeds to a peak of 54Mbps, and was released in 2003.

- **802.11n** – The 802.11n standard (finalized in 2009) upped the maximum raw Wi-Fi connection speed to 150Mbps, and 4-stream MIMO allowed compatible devices to achieve 600Mbps. The 802.11n standard supports both 2.4GHz and 5GHz frequency bands, but 5GHz support is optional, and many 802.11n routers only support 2.4GHz. And some that offer support for 5GHz only allow one frequency band to be used at a time – meaning that enabling 5GHz means locking out all your 2.4GHz equipment!

- **802.11ac** – The 802.11ac standard (finalized in 2013) provides for Wi-Fi speeds well over 1Gbps on 5Ghz. Most 802.11ac gear simultaneously supports 802.11n on 2.4GHz for backwards compatibility too.

- **802.11ad** – Also known as WiGig, the new 802.11ad standard is intended to provide extremely high multi-Gbps speeds using high frequency 60GHz spectrum, with range limited to a single room. This will enable things like wireless hard drives, and ultra-HD video streaming. WiGig-compatible products have been slow to come to market so far.

- **802.11ah** – "HaLow" is the friendly name for the new 802.11ah protocol, a new variant of Wi-Fi that uses the unlicensed 900MHz spectrum that older generation cordless phones operate on. This low-frequency spectrum can travel further through the air, and is much better at penetrating walls and other obstructions than the high-frequency 2.4GHz & 5GHz bands. HaLow has generated excitement because it is intended to double Wi-Fi range, but HaLow is NOT intended for speed. It is primarily focused on connecting basic devices like thermostats and garage door openers, but it may eventually provide for slower speed connections that can cover an entire RV park. The 802.11ah standard was just finalized in early 2016, and the first HaLow products will likely not come to market before 2017 or even 2018.

Range Considerations:
Lower frequencies travel further and are better at penetrating walls, so while 5GHz 802.11ac is the ultimate local wireless network within your RV, it is not well suited for talking to distant campground access points.

For the time being – 2.4GHz 802.11n will remain the technology best suited to longer range Wi-Fi.

Purchasing Tips: If you are buying a Wi-Fi router, look for 802.11n (or 802.11ac), explicit mention of 5GHz, and especially "simultaneous dual band" support, which allows both frequency bands to be used at once.

Unfortunately, 5GHz Wi-Fi support remains rarely found in most cellular routers (like WiFiRanger, MoFi, and Pepwave), though some personal hotspots (such as the Netgear AC791L) have begun to include it.

If you are buying a laptop or tablet, to make sure that you are able to to take advantage of faster speeds in the future – make sure that 5GHz 802.11n or 802.11ac is supported.

For years now, all Apple laptops and iPads have had dual-band support, but the iPhone only added support for 5GHz as of the iPhone 5, and 802.11ac support as of the iPhone 6.

A lot of flagship Android devices already support 802.11ac, but older and cheaper phones, tablets, and laptops are often still 2.4GHz only.

Security on Public Wi-Fi Networks

The internet can be a scary place – it is important to be careful out there!

On a public Wi-Fi network, you should assume that anyone else connected to the same network as you has the potential to eavesdrop on your communications.

You may not think that any of the other campers nearby appear to be hackers, but... have you considered the possibility that some of them may have been hacked in the past?

If a computer has been infected by malware, it may have stealthy spy software running looking for other computers to infect or to log potentially valuable data. The friendly grandma and grandpa with the cute doggie camped at the next spot over may not care that you're online paying your bills, but their obsolete Windows XP laptop might be working for the Russian Mafia – logging everything that happens nearby.

The only sure way to protect all of your communications is to use a VPN (Virtual Private Network) service that encrypts everything between your computer and the VPN provider's central office, bypassing any potential eavesdroppers nearby.

But – is it worth the trouble, decreased speed, and cost to encrypt reading the day's news and checking the weather? Much of what most people do online isn't worth protecting.

A lot of websites are secured using SSL (Secure Socket Layer), which acts like a VPN between your browser and that particular site, effectively blocking anyone from listening in without the computational resources of the NSA to devote to decryption.

In other words, if you load a website using a URL starting with "https://" or your browser displays a padlock symbol – you are generally secure logging into that site, even on public networks.

Almost all financial sites use https, and – after some very embarrassing hacks – even social networks like Facebook and Twitter have begun to default to "https://" connections. Even Google defaults to SSL now.

Always use these secure connections if a site gives you the option!

But very few blogs or online forums give you the option of using https://, as it significantly increases the cost to run the site to do so.

Your biggest actual risk when using a public Wi-Fi network is not when you log into a secure site like your bank, but when you log into insecure websites and online forums using the same username and password combinations that you have also used elsewhere. In this case, it is all too easy for other machines on the same public Wi-Fi network to "sniff" your password and login credentials.

A hacker isn't likely to want to impersonate you on an RVing forum, but he or she will take that password and use that to try and get into other critical secure sites. If you have used the same password and username anywhere else online, your identity is at risk!

> The only way to protect yourself is to pay attention to the number one rule of Internet security – **NEVER ever ever use the same password twice. DON'T DO IT!**

It might seem ridiculous to try and memorize different passwords for every site you visit, but you don't have to. Use a password management program to do it for you. There are several great options out there – we personally use 1Password (agilebits.com/onepassword), which syncs all of our password files between our laptops and mobile devices.

LastPass (https://lastpass.com) also has a great reputation, and a lot of fans. Recent versions of Windows and MacOS X also have more basic password manager capabilities built in too.

If you don't use a password manager – at the very least, keep your key passwords written down in a private paper journal – with a photocopy saved somewhere safe in case of disaster.

Hackers don't like physically breaking in to get things – they can attempt billions of logins an hour (really!) on a basic home PC with some fancy cracking tools, but physically breaking in someplace is messy and time consuming with high risks.

The odds are a lot higher that the RV next door has a laptop listening in for passwords sent to insecure sites than it is that a shady hacker will physically break into your RV to find your passwords hidden deep in your glovebox.

> **Security Update:** If you are still running Windows XP, stop reading this book right now and upgrade (or donate) that machine ASAP.
>
> XP has reached the point where it is no longer being supported with security updates from Microsoft, and the core security in XP is long obsolete. At the same time, the Russian Mafia is still very actively supporting Windows XP, and hacker tools for exploiting XP systems are thriving.
>
> By taking any remaining XP machines off the internet, you are doing not only yourself – but everyone – a favor.

VPN Services – Protection, Privacy, and More

Virtual private networks (aka VPNs) encrypt all the data between your devices and a central VPN server – so that no one on your local network can tell where you are surfing.

In the past, VPNs used to be strictly for advanced users who knew how to manage technical configurations and who had access to their own central server to act as a host.

Now, however, there is a wide range of VPN providers offering incredibly simple-to-use services with clients for Mac, PC, iOS, and Android. No advanced configuration required – just click "On".

Some highly regarded easy-to-use VPN services to consider are:

- Spotflux (www.spotflux.com)

- Cloak (www.getcloak.com)

- TunnelBear (www.tunnelbear.com)

- WiFiRanger "Safe Surf" - Built into WiFiRanger routers.

And some geekier, more advanced options:

- Private Internet Access (www.privateinternetaccess.com)

- AirVPN (www.airvpn.org)

Some VPN services do a lot more than just encrypt your traffic. One of the coolest features many offer is the ability to change the central VPN server you are communicating through – in essence, changing where you appear to be on the internet.

This location-changing feature allows you to appear to be connected from a different country – allowing you to stream BBC shows as if you were in the

UK, for example. Or if you are traveling outside the US, with the help of a VPN, you can still get access to Netflix as if you were still at home.

Some VPN services also offer additional connection optimizations, such as ad blocking or automatic data compression. Shop around and try a few free trials to decide what most appeals to you. VPN services usually offer limited free service plans, or cost $30–$120 a year to secure all of your devices.

Watch What You Share

One more critical and often overlooked security tip for connecting to public Wi-Fi: Be aware of what you're sharing publicly on your computer to other users on the local network!

Once on a RV park's public Wi-Fi network we were able to see and access photos of our neighbor's dog that she was sharing publicly in iPhoto. Obviously, that's probably not a big deal – her dog was cute and brought us a smile. But what if she had been inadvertently sharing a photo library more...umm...personal?

We've discovered people's shared music libraries, movies, and documents.

Know how to turn public sharing on and off in your operating system and key applications, or you might risk sharing something much more embarrassing and personal than your Taylor Swift music library!

If you are not sure what you are sharing, ask a trusted friend on the same network to look for your computer and try to connect. If they can't even see you, you are golden.

And have you ever considered what information your phone or hotspot is giving away about you? Even a password-protected network might reveal way more than you'd imagine – just by what it is named.

We recently had some campground neighbors who's personal hotspot name was "John Doe's iPhone" (name changed to protect the clueless). His network was password protected and secure, but his name is rather unique. Two seconds in Google reveals that he is a 55–59-year-old male, and we even got his home address and phone number.

The lesson: If you want to remain anonymous to your neighbors, take care what you name your network, and don't rely on the default auto-generated name your phone may use when creating a hotspot.

Or, have you fun with this. Come up with clever network names like "FBI Surveillance Van," "Have a nice day :)," or even broadcast your travel blog with something like "technomadia.com" or "Bring Wine to Site 29 for Happy Hour" as a network name as a way to make new friends!

Wi-Fi Range Extending Gear

Ok – so you found a Wi-Fi hotspot to use – but you can't get the signal while sitting in your RV.

How do you get online via Wi-Fi, without needing to spend your days sitting at Starbucks and your nights lounging on the RV park's office porch?

Wi-Fi was never intended to be used for long range networking, but it is often possible to push the limits of what Wi-Fi is capable of.

But first – do yourself a favor and test to make sure that it will actually be worth the effort!

The Wi-Fi Worthiness Test

A lot of people invest a small fortune in long-range Wi-Fi hardware, only to report back disappointedly that it hardly made any difference.

In a lot of these cases – there just wasn't any worthwhile signal to work with in the first place. If the campground has slow and unreliable Wi-Fi in the front office or rec center near the hotspot, no amount of technology will be able to make things any better than that back at your rig.

Before you invest time and money in getting connected via Wi-Fi, find out if the hotspot you're trying to connect to is actually worth the effort.

Do this by taking your laptop, phone, or tablet up as close to the hotspot as you can manage – and then run some speed tests. Try out some typical web surfing. Maybe even stream a video.

If the experience is a good one, then using long range Wi-Fi gear in your RV may benefit you. If not – save yourself some frustration and find another way online.

Wi-Fi Range Capabilities

Generally, there is a rough hierarchy of Wi-Fi range capability – from the shortest range gear to the longest:

1) **Wi-Fi Gadgets** (including phones, tablets, etc)

2) **Laptops**

3) **Indoor WiFi-as-WAN Routers** (WiFiRanger Go, Pepwave Surf SOHO, etc.)

4) **High-Power USB Wi-Fi Network Adapters** (Alfa, etc.)

5) **Outdoor CPE / Router** (WiFiRanger Sky, WiFiRanger Elite, Wave WiFi Rogue Wave, The Wirie AP+, etc.)

6) **Outdoor CPE with Directional Antenna** (Ubiquiti NanoStation, parabolic dish, etc.)

Though nearly every laptop and mobile gadget now comes with Wi-Fi capabilities built in, the integrated systems on most aren't engineered with antennas and transmitters designed for connecting to a hotspot far away.

Very few laptops or mobile devices make any provision for using an external Wi-Fi antenna either, so other than by balancing your laptop in a window, there is no way to increase your Wi-Fi range without adding another external device to the mix.

Fortunately – there are options to increase your range. It all depends on how much cost and complexity you are willing to add to your connectivity arsenal.

Read on and we'll explain all of the more advanced options.

And for some additional tips on things that can impact Wi-Fi connections, see the chapter on "Wireless Signal Enhancing Tips."

WiFi-as-WAN & Wi-Fi Repeating

A router connects to an upstream "wide area network" (WAN, otherwise known as "the internet"), and provides service to multiple downstream devices on the local area network (LAN, in other words, your devices) by routing where the network traffic goes.

A home router is usually plugged into a cable or DSL modem for the upstream WAN connection, and provides a local Wi-Fi LAN network for you to connect to.

Wi-Fi Range Extending Gear

On the road however, it's going to be rare for you to have DSL or cable to use as your internet source – and you may want to use a distant Wi-Fi network instead.

Some Wi-Fi routers support this – using another Wi-Fi network as the upstream connection. They connect to the shared WAN (for example, the campground's Wi-Fi hotspot) and then distribute the connection to your local devices as your own private wireless LAN.

This feature is called WiFi-as-WAN.

This is a rare feature in typical off-the-shelf home routers, but is common in mobile & travel routers. WiFiRanger, Pepwave, and Cradlepoint are all examples of mobile router brands that support WiFi-as-WAN as a core feature.

Since a WiFi-as-WAN router will usually have a stronger radio and better antennas than any of your laptops or tablets, a WiFi-as-WAN router can act as a relay. This lets devices on your LAN share access to a Wi-Fi network further away than they ever could reach alone.

Routers and WiFi-as-WAN are discussed further in the Routers chapter later in this book.

But even the best WiFi-as-WAN router is limited by where you can place it inside your RV. If it is stuck in a lower enclosed cabinet, or your RV is built with metal walls, the router may actually perform worse than your laptop sitting on an exposed table near a window is capable of.

USB Wi-Fi Network Boosters

When it comes to Wi-Fi range – nothing beats having a direct line of sight between the access point and your device with as few obstructions as possible.

If you have a window that faces towards the Wi-Fi hotspot you are trying to connect to, one of the most affordable ways to get a bit more range is to use a USB Wi-Fi network adapter with a more powerful Wi-Fi radio.

These are often marketed as "antennas" or "boosters", but if they are designed to connect to a single computer via USB they are actually fully self-contained external Wi-Fi networking cards. Similar to what is built into your computer, only with a much more powerful radio transmitter and a more capable attached antenna.

Wi-Fi Range Extending Gear

With the included extension cables, you can plug into your laptop and set these small devices in a window facing towards the target hotspot. A few of these devices are even designed to be used outdoors, connected to your computer with a 30' long USB extension cable to give you the ability to reach up to the roof.

Some of the better know manufacturers of these devices include:

- Alfa Network – (www.alfa.com.tw)

- Ideaworks – Found in some retailers like Walmart and Sears.

- C. Crane – (www.ccrane.com)

- BearExtender – (bearextender.com)

The downside of USB adapters like these is that they will only get one single computer online, and they cannot help connect your non-USB devices to the distant Wi-Fi network unless you also buy a special companion router (like the Alfa WiFi USB Repeater) to work with your chosen USB booster.

A lot of these devices have been slow to release driver updates to support the latest Mac OS and Windows releases, so be careful to double check compatibility.

Roof Mounted & Directional CPEs

Particularly with campground Wi-Fi, a good signal is often very hard to find at ground level. If you want to connect over the greatest possible distance, nothing beats having a permanent long-range Wi-Fi setup on your roof.

But unfortunately, Wi-Fi signals degrade rapidly when traveling over an antenna cable. Though it is possible to use an outdoor antenna connected to an indoor Wi-Fi radio via a coax cable, the amount of loss makes this rarely a good idea.

Instead of having the electronics indoors and the antenna outdoors – it is actually very common to mate the electronic brain of the Wi-Fi radio directly with the antenna in a single package rated for outdoor use.

Wi-Fi Range Extending Gear

This is essentially what a CPE is.

CPE stands for "customer premises equipment" the common term used to describe the commercial-grade outdoor-rated Wi-Fi access points used by wireless service providers. A CPE combines roof-mounted height, better antennas, and a more powerful transmitter to deliver the maximum possible Wi-Fi range.

Odds are – any campground with a professionally installed Wi-Fi network has CPE hardware mounted on poles or buildings around the park. You can use this same caliber of hardware on your RV to ensure the best possible long range connection.

If you are comfortable configuring a device designed for network engineers and not average consumers, you can use a commercial roof-mounted CPE paired with any indoor router of your choosing.

Key manufacturers include:

- **Ubiquiti** (www.ubnt.com)

- **MicroTik** (www.mikrotik.com),

- **EnGenius** (www.engeniusnetworks.com)

If you don't mind needing to hunt for signals and manually aim at each location you set up, you can even use a CPE model with a built in directional antenna – like the Ubiquiti NanoStation 2.

The downside of commercial-grade gear like this is that you can't count on any customer support or handholding for RV installations, and the software configuration can be overwhelming.

For a much simpler installation, you can instead go with a CPE that has been packaged and optimized for the RV or marine market. Check out these manufacturers:

- **WiFiRanger** (www.wifiranger.com) – The roof-mounted Sky & Elite are very popular with RVers, and work hand-in-hand with an indoor WiFiRanger router to build an integrated system.

- **Wave WiFi** (www.wavewifi.com) – The Rogue Wave product line lacks many of the advanced features of the WiFiRanger, but has a good reputation overall.

- **The Wirie** (www.thewirie.com) – Primarily focused on the marine market, the AP+ is at home mounted on a mast or an RV's ladder.

These devices are actually commercial-grade CPE hardware under the hood (often with guts actually manufactured by Ubiquiti or MicroTik) with a simplified consumer-friendly user interface on top.

All of these devices are designed to be installed permanently outdoors directly onto the roof of an RV or boat, maximizing range.

Having a roof-mounted Wi-Fi access point paired with an indoor Wi-Fi router is especially handy if your RV is a metal signal-blocking tube, such as a bus conversion or Airstream. Getting the transmitter up on the roof means you don't have to fidget with your gadgets trying to figure out which window has the best signal today.

The upfront cost for going with a roof-mounted CPE setup may be steeper, but this is where the most substantial range and speed gains are to be had. If long range Wi-Fi is central to your connectivity needs, it is often worth investing in doing it right.

Keep Current Alert

We are constantly tracking and testing routers, antennas and gear. Be sure to check our Resource Center at www.rvmobileinternet.com/resources for any current reviews or product comparisons we might have available.

Wi-Fi Is A Two Way Street

No matter how much you invest in commercial-grade high quality CPE hardware mounted on the roof of your RV, the quality of the hotspot you are connecting to is usually outside your control.

While some campgrounds will have also invested in long range outdoor CPE gear, many others have little more than an old obsolete bargain-bin home wireless router from Walmart sitting behind the desk in the office.

And that caliber of hardware is never going to be able to communicate over any substantial distance, no matter what you do on your end.

But sometimes you can take matters into your own hands – upgrading the hotspot you are communicating with to vastly increase your range. Particularly when you are driveway surfing with friends and relatives, it can be useful to set up a temporary high-power wireless access point in a window that is connected by ethernet to their home network.

And some RVers have even volunteered to upgrade their favorite campground's Wi-Fi equipment, purely for the selfish reason of getting a better signal at their site.

Wireless Signal Enhancing Tips

There is not much that is more frustrating than the agony of having just one bar of signal, whether that signal happens to be Wi-Fi or cellular.

That solitary bar is a cruel tease – usually not enough signal to actually reliably use, but it's there, taunting you.

If you maybe just hit reload one more time… or shift your position a bit…

Right at the moment you are about to give up, it works! For a few minutes, at least. Just long enough to keep you on the hook trying.

At least when there is no signal at all, you can concentrate on doing other offline things. But having a hint of signal…that is the path to madness and not accomplishing anything.

There are, however, things you can do to improve a bad situation – taking a weak signal and making the most with it. With external antennas, cellular boosters, and Wi-Fi repeaters, the results can sometimes be near miraculous.

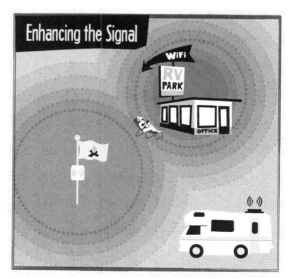

But other times, no amount of boosting can help – that cruel solitary bar might remain undefeated, taunting you still.

Things That Impact Wireless Signals

Wireless signals can be quite variable and are influenced by so many things. Sometimes just moving from one side of your rig to the other can make a dramatic difference.

In general, there are a few universal things that help with reception, and these guidelines are relevant for both cellular and Wi-Fi.

Things that help:

- **Line of Sight:** Nothing improves a wireless signal more than having nothing between the sender and receiver. If you can visually see the cell tower or the Wi-Fi hotspot, there is a good chance your wireless device can too.

- **Altitude:** The best way to get line of sight, or at least fewer obstructions, is to get up over the clutter. An antenna at the top of a 20' mast has nothing but clear air around it. An antenna mounted directly on the roof might be blocked when a giant fifth-wheel moves in next door, or even by your own air conditioner. And an antenna stuffed into the back of a crowded lower cabinet is starting off with a substantial handicap.

- **Directional Focus:** A transmitter of a given loudness (power) can be heard a lot farther away if the energy is focused towards the receiver – just as speaking into a megaphone does wonders to project your voice. Of course, the downside of having a directional antenna like this is that it adds a manual aiming step. Since the signal off to the sides outside of the sweet spot is substantially diminished, directional antennas are often more trouble than they are worth if you're just passing through. Only in a relatively fixed location is the extra aiming effort worthwhile.

- **Power:** Louder is better – but only up to a point. If you push too much power out a radio antenna, it can begin to overwhelm your target. And worse, it can drown out others using the same or nearby frequencies.

- **A Sensitive Listener:** Just like in relationships, often better than raw power is a sensitive ear. If the receiver is listening carefully, even a weak signal might be heard. This sensitivity is often a prime difference between expensive commercial-grade wireless gear and cheap consumer-grade stuff.

- **An Uncrowded Channel:** Wireless networking devices are designed to allow multiple users to share a single channel. On a cell tower or a crowded campground Wi-Fi network, there could be hundreds of devices all trying to talk at once. The more people talking at once, the more congested and degraded the network becomes for everyone.

- **Upstream Capacity:** A river can't flow through a straw – and often the real problem with both campground Wi-Fi and cell towers is that there's only a straw's worth of pipe bringing in the water. Some campgrounds invest in great Wi-Fi equipment, and you might have the strongest signal ever – but upstream they have skimped and have little more than a single basic DSL line serving the entire campground. The same applies to cell towers, particularly in more remote areas. The tower may speak fast LTE, but if the upstream network (called the backhaul) is not sufficient for the population being serviced, the actual speed you see may be extremely limited.

- **Antenna Diversity:** Having multiple antennas working together is called antenna diversity, and this can work wonders with compatible equipment – particularly in areas where there may be a lot of signal reflections bouncing around. A secondary antenna can compensate for a primary antenna being in signal shadow.

There are some universal things that degrade all signals – both cellular and Wi-Fi. Things that hurt:

- **Metal Obstructions:** Barriers between you and the Wi-Fi base station or cell tower are bad. But metal barriers are very, very bad. Radio waves in general have difficulty passing through metal, making getting a signal for those living inside metal buses or shiny Airstreams an extreme challenge.

- **Interference:** The more noise on the line, the harder it is for a radio receiver to hear. Noise can originate with other intentional signals or it could be background noise from microwave ovens, hair dryers, or even the sun.

- **Overcrowding:** Overcrowding leads to a vicious spiral – with too many radios trying to broadcast on one channel, it becomes hard to pick out individual conversations. Transmitters try to compensate by broadcasting louder and repeating themselves to get a message through. That leads to even more overcrowding and noise – until eventually the wireless network saturates and barely any data is flowing.

- **Power:** Some Wi-Fi devices and routers let you manually set your radio power. It seems counterintuitive, but you can actually sometimes improve both speed and even range by dropping down to a lower power setting. At full power, you may be too loud, and you could be generating interference and overcrowding your neighbors. Speaking in a whisper is sometimes more effective than roaring at the top of your lungs.

A lot of these things may be out of your control – but it helps to understand the things that may be impacting your signal in any given situation.

Understanding MIMO

MIMO (Multiple Input, Multiple Output) is one of the core technologies enabling LTE cellular, and almost every LTE mobile device (whether a phone or a hotspot) actually has TWO cellular antennas on board to enable the magic of MIMO.

On the other end of the line – LTE cell towers typically have two or four or even eight antennas working together in tight synchronization to communicate with you.

With more antennas transmitting a signal, there are more possible echoes and reflections for the receiving device to extract a signal from.

And more reception antennas on the cell tower better enable your carrier to receive a weak remote signal from your handset.

This figure illustrates a 4x2 MIMO deployment – with four antennas on the tower communicating with two antennas inside a cellular hotspot:

Depending on the conditions, the LTE tower and receiver will negotiate one of several possible MIMO modes. These are the important ones to know about:

- **Mode 1** – This is a fallback mode using a single transmit and receive antenna, essentially disabling MIMO.

- **Mode 2 "Transmit Diversity"** – This is the default mode for LTE, and it calls for the tower to transmit the same data stream over multiple antennas, just encoded differently. This does not result in any increase in speeds, but the redundancy makes for a more reliable signal, especially in areas where the signal is weak and there are not a lot of "echoes" – such as rural areas without many buildings to reflect a signal off of.

- **Mode 3 "Spatial Multiplexing"** – This is the turbo-mode, where different data streams are transmitted over each antenna, and are then combined at the receiver for a 2x speed boost.

In urban areas with a lot of signal reflections, MIMO's spatial multiplexing mode can almost magically double cellular speeds.

And in rural areas, transmit diversity allows for LTE devices to work wonders with weak signals – often delivering seemingly impossible speeds when the signal strength is less than -100dB, the sort of signal that would have been barely useable on 3G or 4G networks.

MIMO & Boosters Often Do Not Mix

A cellular booster works by taking a signal picked up by a single external antenna, amplifying it, and then rebroadcasting it on a single interior antenna.

Funneling your cellular signal through a single antenna and amplifier means that all the advantages of MIMO are being thrown out the window, and the two antennas on your phone or hotspot are now picking up essentially the same thing.

When confronted with a boosted signal, LTE modems often have no choice but to fall back to the non-MIMO Mode 1.

An Analogy: Think of using a booster akin to plugging one ear, and sticking a powerful hearing aide into the other. With amplification, you will be able to focus and hear a conversation from much further away than you would otherwise, but you will be giving up your stereo hearing and all the benefits that come from having two ears and being able to listen to two things at once.

The booster also acts as a megaphone for your transmissions – making it a lot easier for you to be heard from further away. But being able to transmit louder comes with a price – losing MIMO.

Cellular boosting is all about tradeoffs – in the weakest signal areas a booster can get you online where you otherwise wouldn't have any hope of connecting at all.

But in strong signal areas, turning on a cellular booster can be the equivalent of slamming on the speed brakes.

With a booster you will often see "more bars" despite the halved speeds.

With MIMO – signal strength is clearly not the whole story.

Where exactly the dividing line sits between a booster helping and hurting you will vary substantially as you travel around – and the only way to know for sure is to test in each location, and with each carrier.

In many cases – using a cellular booster may end up cutting your download speeds in half, while at the same time more than doubling your upload speeds.

You will have to decide what tradeoff to make for each location.

External MIMO Antennas

Some cellular devices support dual antenna inputs – so it is possible to use external MIMO antennas to get the advantage of roof-mounted line of sight without giving up the benefits of MIMO.

Hotspots like the Netgear AC791L, AT&T Unite Pro, Pantech MHS291L, or cellular routers like the Pepwave MAX BR1 all support external MIMO antennas.

MIMO – Not Just For LTE

The powerful multi-antenna techniques that MIMO enables for LTE have also been incorporated into the 802.11n and 802.11ac Wi-Fi standards – allowing for multiple antennas to be put to use to turbo-charge Wi-Fi speeds between compatible devices.

The latest high-end routers, laptops, and desktops support up to 3x3 802.11ac MIMO – using three transmit and three receive antennas working together to give peak Wi-Fi data speeds up to 1.3 Gbps, and beyond.

Even the latest handhelds like the iPhone 6S support 2x2 MIMO 802.11ac – giving a theoretical Wi-Fi speed of 867Mbps.

Keep Realistic Expectations

A lot of people invest big dollars into antennas, cellular boosters, and Wi-Fi repeaters expecting miracles. And in the end, many end up being disappointed when their lofty expectations are not met.

Antennas and boosters can only do so much. If there's nothing to boost, no amount of signal amplification can make something out of nothing.

And even if you get a stronger signal thanks to a booster – if the real speed bottleneck is located upstream, then you might not actually see any practical improvement anyway, at least when it comes to your online experience.

But if you keep realistic expectations, a cellular booster and a Wi-Fi repeater can grow to become some of the most essential elements of your tech arsenal. We've lost count of how many places we've been that would not have been doable without a booster on board.

When it works, it becomes absolutely indispensable.

Antenna Selection & Installation

by guest author Jack Mayer

Antennas are key to successful mobile internet, for both Wi-Fi and cellular.

The antenna you select, or that the manufacturer selects, has more influence on your internet experience than almost any other single piece of your equipment. But for most people, antennas are mysterious devices built into their tech or sitting forgotten on their roof.

How do you tell if the antenna you are using is the optimal choice?

In this chapter, we take a look at antenna types, some of the terms and technology used, how to select an appropriate antenna, and how to mount it once you own it.

Although antennas are an important part of any connectivity solution, they are dependent on clear line of sight, or near clear line of sight, to the signal source.

Getting the antenna up high, away from obstructions, is just as important as the performance characteristics of the antenna itself.

Antennas are an incredibly complex topic if you want to understand all the intricacies of them. People earn PhDs in antenna theory and design. Here we look at the practical aspects of antennas – not antenna theory. We gloss over or take some liberties with some of the intricacies of antenna technology to keep it simple and less boring.

Antenna Performance & Gain

The primary performance attributes used to describe all antennas and amplifiers is gain.

So what is gain, and how does it affect your antenna choice?

Gain is a measurement of the ability of an antenna to focus power on a target location, or the increase in the amount of power brought about by passing a signal through a booster.

If you think of a traditional bare light bulb, the light emanates in all directions. If you then think of a flashlight, that same bulb has the light focused with a reflector so that it is more directional, and thus appears brighter. And if you increase the wattage of the bulb in the flashlight so that you can see it from even farther away, that is amplification and represents a further gain in transmission power.

In this flashlight example, the overall gain would be the amount the perceived brightness changed – a combination of better directionality provided by an antenna and increased power by a boosting amplifier.

Gain is always measured relative to something. In an antenna, the measure of gain is sometimes expressed in dBi . This is decibels relative to a specific type of theoretical reference antenna – an isotropic radiator – that radiates equally in all directions.

A higher gain means that more power reaches the end point of the transmission.

You might also see gain specified as dB (not dBi). Gain measured as dB is always less than the dBi measurements, and is a more realistic measure of relative performance. Figure on around a 2 dB loss when converting from dBi to dB. In other words if a manufacturer specifies an 8 dBi gain on an antenna, expect an actual gain of around 6 dB in use.

Make sure that when comparing specifications on antennas that you are comparing "apples to apples" (dB to dB, not dBi to dB).

Gain is logarithmic. Without getting into the math, the simple rule of thumb is that for each 3dB of gain, the measured power at the receiver is doubled.

Gain in an antenna is often highly misunderstood – but it is simply the amount of focusing of the signal. A 0 dBi gain antenna radiates energy in all directions equally, while a 5 dBi gain antenna focuses the energy in certain directions more than others – but the total amount of energy radiated remains the same.

The gain of a given antenna is measured at the point of the peak energy concentration.

Related to gain is beam width. Beam width is a measure of the amount of focus the radio signal has. The narrower the beam width is, the more focused the antenna and the higher the gain.

Antenna Selection & Installation

Beam widths are generally stated in degrees of horizontal and vertical focus. The horizontal beam width describes where the signal radiates outward from the antenna. An omnidirectional antenna has a 360-degree horizontal beam, since it is radiating in all directions around it. However, it will have a vertical beam width that varies. The vertical beam width describes the angle of the signal radiating outward, relative to the ground.

To keep things simple and more understandable, we are going to take the liberty of categorizing the typical antennas used for cellular and Wi-Fi into two major categories based on their signal radiation characteristics: omnidirectional (omni, for short) and directional.

With an omnidirectional antenna, a higher dBi rating generally means a tighter vertical beam, since the energy gets focused more in a plane parallel to the ground.

For example, an 8 dBi omni antenna may have a 15-degree vertical beam, while a 15 dBi omni may have a narrower 6-degree vertical beam.

The pattern from an omni antenna can generally be thought of in terms of a donut shape, where the antenna is the center of the donut and the shape of the donut represents the signal radiation. The signal is focused outward from the antenna in the horizontal plane – so directly above and below the antenna there is (relatively) little signal.

Thus a lower gain antenna may actually outperform a high-gain antenna if the signal is bouncing among hills, mountains, or city structures – or if the tower is located very high relative to the receiver.

A high-gain omnidirectional antenna, on the other hand, is most useful when towers and receivers have a clear line of sight to each other parallel to the ground – in other words, when the tower is farther away and on the horizon.

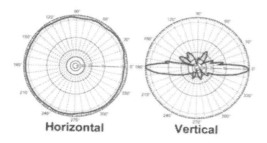

Horizontal **Vertical**

This diagram shows an omnidirectional antenna radiation pattern. In the horizontal plane – looking down on the antenna from above (the antenna is in the center) the antenna radiates equivalently in 360-degrees.

The vertical diagram represents a "side view" of the omnidirectional signal. You can see that there are some "lobes" of signal that radiate upward, and others (smaller) that radiate downward. But most of the signal is sent directly away from the antenna parallel to the ground.

113

The vertical beam is important in omnidirectional antennas – higher dBi omnis have very tight vertical beams, so if they are used close to a receiving station they may "miss" with the main part of the signal.

Also – because the signal travels parallel to the ground, it is important for omnidirectional antennas to be mounted vertically.

While it might not need to be specifically aimed, an omnidirectional antenna mounted horizontally will deliver very disappointing results.

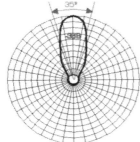

This diagram is a view of the signal from a directional antenna, looking down on the antenna from above. The signal radiates outward from the antenna in the center – the beam width in this case is 35 degrees – a sharp contrast to the omnidirectional antenna above.

You can see that the signal is focused forward in a single direction – making aiming critical.

Choosing an Antenna

Using the best available equipment and directional antennas, you can often expect to get usable Wi-Fi at up to a half to a full mile away from a quality base station, assuming a clear line of sight to the station.

In some circumstances – like across water – you can capture Wi-Fi from much farther (as much as two+ miles), but typically the limitation is the power of the access point. Wi-Fi hotspots (access points) are not typically designed for long-distance networks, so the hotspot may not have enough signal to reach your location, even though you have enough power to reach it.

Cellular is very dependent on how the tower you are accessing is configured – some towers have distance constraints engineered into them. It also depends on the frequencies and protocols being used. But it is fair to say that 18–20 miles is generally the best you will see for cellular communication, no matter what antenna and booster you use.

Every antenna is designed and optimized to operate over given frequency ranges.

For Wi-Fi, there are two frequency bands: the 2.4 GHz and 5 GHz ranges.

Cellular frequencies, on the other hand, vary by carrier and location (see the chapter "Understanding Cellular Frequencies"), so cellular antennas are challenged to pick up a wider range of frequencies than are Wi-Fi antennas – and are often only optimized for a few.

When looking at cellular antennas, you must make certain that the antenna is designed for the frequency bands your carrier is using.

Generally cellular and Wi-Fi antennas are not interchangeable.

Antennas should have a spec sheet that lists the gain at various different frequencies – use this to confirm how well the antenna will perform with your carriers.

Antennas also come in various types or form factors.

Most people are familiar with the antennas that are seen mounted on vehicles. In the old days these were simple metal shafts. But for high-performance computer and cellular applications, there are more specialized antennas.

The two basic types of antennas – omnidirectional and directional – are available supporting both cellular and Wi-Fi frequencies.

Omnidirectional Antennas

These are the easiest antennas to use, since they require no aiming. But radiating energy in 360 degrees has the disadvantage of diluting the signal. The receiving station – be it a Wi-Fi access point or a cellular tower – is in a fixed location, thus most of the signal energy is wasted in transit to it.

Omnidirectional antennas are also more prone to outside RF (radio frequency) interference – they are, after all, receiving both signal and noise from all directions.

For an antenna in motion, omnidirectional capability is a requirement. You have no idea where the receiving station is so you have to broadcast in all directions in order to ensure that you can hit the receiver.

Most mobile users will benefit from the passive ease of omnidirectional antennas. If you pull into an RV park, you will want to access the RV park's access point (AP) to receive Wi-Fi without needing to aim an external antenna. An omni antenna does this quite well. But because some of the energy is wasted by not being focused on the access point, performance may suffer.

The same thing applies to cellular voice and data reception – you typically want access without needing to aim an antenna first.

There are many form factors of omni antennas. Fixed-mount antennas use a bracket mounted to the side of an RV, vehicle, or building, or they may actually be mounted through the roof of a vehicle.

Magnetic mount antennas use a magnet to hold the antenna to a metal surface – typically the roof of a vehicle. But any metal surface can be used – for example, a metal plate held to the roof of an RV with caulk.

Glass-mounted omnidirectional antennas offer a mounting method that can place the antenna high on a vehicle and not damage the body. Glass-mount antennas pass the signal through the glass without needing a hole: The interior patch with the lead is placed directly under the exterior patch that holds the actual antenna. The glass acts as a capacitor, but allows the AC energy of the signal to pass through to the antenna. It is critical for the glass to be nonmetallic – otherwise, the signal cannot be passed. Solar glass or glass with a defrosting grid in it can cause signal issues, so you have to be sure of the type of glass you are dealing with. On an RV, dual-pane glass on windows will not work properly with this type of antenna.

Glass-mount antennas are commonly used on vehicles for cellular frequencies. They can work well if properly installed and tuned. For use on an RV, a higher performance antenna is often preferred, but a glass-mount antenna on the windshield of a motor home can perform adequately if no other option is available.

Some multi-band antennas combine multiple internal antennas into a single casing – with a separate antenna wire for each device to be hooked up. This is the only way that a single antenna should be used to connect to multiple devices simultaneously. A single antenna wire should never be used with a splitter to plug into multiple devices.

Examples of Omnidirectional Antennas

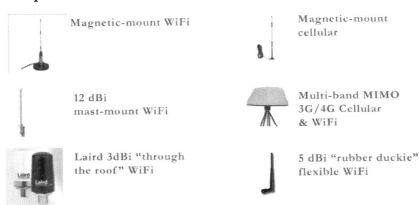

Magnetic-mount WiFi

Magnetic-mount cellular

12 dBi mast-mount WiFi

Multi-band MIMO 3G/4G Cellular & WiFi

Laird 3dBi "through the roof" WiFi

5 dBi "rubber duckie" flexible WiFi

Directional Antennas

With a directional antenna, all of the radio energy is focused into a single direction, with the horizontal beam width generally 120 degrees or less. The vertical beam will also be more focused.

How tight the beam is focused varies by antenna – but it can be as narrow as 10 degrees, or even less.

If you think of an omnidirectional antenna as a lantern, you can imagine a directional antenna as a flashlight. The more powerful the antenna, the tighter the beam is focused, and the more critical aim becomes.

Directional antennas used for cellular and Wi-Fi fall into broad categories.

The widest beam antennas are generally called sectional antennas. These offer wide coverage areas – generally around 120 degrees, so three sectional antennas on a mast working together can cover 360 degrees, and together offer more focused coverage than a single omni antenna could. This setup is commonly seen on cell towers. It is also used for Wi-Fi where only part of an RV park needs coverage.

Next in coverage is a panel antenna. The panel antenna is more focused than the sectional and is available in a range of beam widths as tight as 20 degrees. Panel antennas are often typically smaller than the sectional antennas and are thus a better selection for most mobile requirements.

Yagi antennas perform similarly to panel antennas, with a tightly focused beam width in both the horizontal and vertical planes. Some Yagi antennas have wider beams and require less precision to aim.

Grid antennas and dish antennas use a parabolic reflector to focus the energy even tighter – generally around 7–10 degrees both horizontally and vertically. There is a feed antenna suspended in front of the grid or dish whose purpose is to transmit/receive signal. The dish acts as a reflector and focuses the signal into a very tight cone. Dish and grid antennas are most commonly used for longer-distance point-to-point connections.

There is some overlap in all of these specifications, but in general the sector antennas have the broadest beam, and the antennas using a parabolic reflector (grid and dish) have the most focused beam.

Grid and dish antennas are much harder to aim than panel antennas.

MIMO Antenna Technology

Multiple-input multiple-output (MIMO) technology uses multiple antennas at both the transmitter and receiver to greatly improve both performance and reliability.

MIMO is part of the LTE cellular standard and the 802.11n and 802.11ac Wi-Fi standards.

Some LTE devices (like the Verizon Netgear Jetpack AC791L on Verizon and the Netgear Unite Pro on AT&T) support dual external antennas, allowing for the combination of MIMO capabilities with the benefits of external antennas.

To better understand what MIMO is and how it works – see the "Understanding MIMO" section of the "Wireless Signal Enhancing Tips" chapter.

Highly Regarded Antenna Sources

- 3G Store (www.3gstore.com)

- Powerful Signal (www.powerfulsignal.com)

- WPSAntennas (www.wpsantennas.com)

Understanding Ground Planes

Many of the mobile boosters that are ideal for RV use come with a short stubby rubber magnetic mounted antenna. These antennas are specifically designed for automobile use, and for being placed on the roof of a metal car or truck.

The magnet keeps the antenna connected to the metal roof, however many do not realize that the metal roof is also an integral part of the antenna design. Most RVs however don't have a metal roof – they are fiberglass, rubber or some other material – and these antennas will end up performing poorly.

Why is a Ground Plane Necessary?

A ground plane is a reflective metal surface that the signal from an antenna bounces off of to better meet the antenna's design goals.

Antennas requiring ground planes are called ground-plane dependent, and having an appropriate ground plane can greatly affect performance. A dependent antenna can become effectively useless without one. You should always ensure that any antenna requiring an external ground plane –

typically magnetic-mount antennas – is placed upon an effective reflective surface.

Antennas that do not require an external ground plane are called ground-plane independent antennas. These antennas have the means to reflect and focus the signal without the external ground plane being present. For most ground-plane independent antennas, having a ground plane present will still somewhat enhance signal gain, but it is not required.

How to Create a Ground Plane on a Non-Metal RV

Most RV roofs are not metal, so you'll have to replicate the metal of the automobile roof that these antennas were designed for. Thankfully, it's a pretty easy modification to make – just provide a piece of metal on your roof, and mount the antenna to that.

The minimum size of a metal ground plane should be one-fourth of the wavelength of the radio signal being broadcast.

For most cellular transmission frequencies, a disc of some sort of metal at least 8" in diameter is generally considered sufficient and actually provides some room for error. For example, many LTE bands are broadcast in the 700 mHZ range, which  has a wavelength of 16.8" – so 16.8 / 4 would be just around 4-5" needed for optimal performance of those bands.

Most magnetic antenna manufacturers will specify the size required.

To be absolutely sure, an 8" diameter plate always works.

The plate does not have to be heavy/thick and it only needs to be magnetic if you want to use the magnet on the antenna base to attach the antenna. You can also use aluminum foil or metal tape in a pinch.

The shape is also not critical: A ground plane can be rectangular or circular, as long as it meets the minimum size measure in all directions. Roof shingle flashing (comes precut to 8" x 12", galvanized steel and usually under a buck at the hardware store!), flat cookie sheets of the correct size, pizza pans, old circular saw blades, paint can lids, and many other things can all make good ground planes for most antennas.

To attach a magnetic antenna to a fiberglass or rubber roof, take your metal plate and spray it with your choice of rust-resistant paint, and then simply attach it to your roof with a compatible caulk.

Then place the antenna in the center of the plate, and let the magnet hold it. And then at the side of the plate, secure the wire to the roof with a puddle of caulk.

The advantage of this method of attachment is that if a tree limb hits the antenna, it will simply flip off the mounting plate and be retained on the roof by the caulk puddle. Later, it is a simple matter to right the antenna back on to the plate.

A rigidly mounted antenna on the other hand is easily damaged by tree limbs, however you can attach the antenna by other means (silicon, adhesive, etc) if your ground plane is not a ferrous metal.

Antenna Cables & Calculating Overall System Gain

Another factor that affects performance of all antennas is signal attenuation due to the cable length, size, and quality between the antenna and the radio.

Attenuation is a general term meaning reduction in the strength of a signal over distance. In the context of antennas used in a mobile environment, it is almost always the result of cable length.

You need to use good quality antenna cables and good connectors or the signal will be compromised, defeating entirely the benefits from using a high-quality external antenna or booster.

The lower the frequency of the signal, the less attenuation there is in a given length of cable. So the cellular LTE 700MHz frequencies show significantly less signal loss over the same length of wire than the 1900MHz PCS or AWS 1700/2100MHz frequencies.

Wi-Fi signals at 2.4GHz degrade rapidly over antenna cables, and 5GHz Wi-Fi is not well suited for even short wire runs between the receiver and the antenna. This is why it often makes a lot of sense to put the entire Wi-Fi radio up on the roof and connect to it digitally over Ethernet, which does not suffer attenuation with distance.

Cable attenuation is expressed in decibels of loss – so it is typical to see a cable extension specified as "attenuates signal 2dB" (if it is expressed at all).

Most cellular antennas come with RG-58 or RG-174 cable, with a 12–15' long section often hardwired to the antenna. These wire types are thin and flexible, but for extension cables and longer runs, heftier cables should be used.

Antenna Selection & Installation

Here are the cable types you are most likely to see in use in an RV:

Cable Type	Attenuation of a 750MHz Signal	Attenuation of a 2.4GHz Signal	Comment
RG-174	23.6dB/100ft	75dB/100ft	Very thin and flexible but even higher loss than RG-58. Common on magnetic roof antennas and interior antennas.
RG-58	13.1dB/100ft	32.2dB/100ft	Commonly comes on trucker and marine antennas. Suitable for extensions of 20' or less.
LMR240	6.9dB/100ft	12.9dB/100ft	An excellent choice for extension runs inside an RV or up an antenna mast.
RG-6	5.6dB/100ft	N/A	Commonly used for cable TV and satellite wires, but due to the 75 ohm impedance should NOT be used for cellular or Wi-Fi antennas unless you are certain your equipment is compatible (rare).
LMR400	3.5dB/100ft	6.8dB/100ft	Appropriate for small buildings, only rarely used in RV applications for long cable runs or tall masts. It is a very heavy cable, and does not flex or bend well.

In general, any cable run over 30' needs to be looked at very critically to manage signal attenuation.

Remember that the dB scale is logarithmic – 6dB of attenuation over a wire run will result in half the power getting through as a 3dB attenuation run.

Every connection point adds roughly 0.5dB attenuation per junction – so it is always better to use a single long cable than to chain two or more cables together.

You should never ever use a splitter with cellular and Wi-Fi antennas – splitting a signal will cut the power received in half, and two transmitters connected to one antenna is a recipe for trouble.

Your goal for any extension cable should be a total attenuation loss of less than 4dB over the distance you want to cover.

Low-loss cables are more expensive and are stiff and harder to work with — all things to keep in mind when planning your antenna cable runs.

To determine your overall system gain at a given frequency, you add the gains provided by the outside (and inside) antennas, subtract the attenuation from the antenna cables and extensions, and add any gain provided by an amplifying booster.

An Antenna Evaluation Case Study

Say we want to mount a typical magnetic mount omnidirectional cellular antenna on our RV roof specifically to improve Verizon 4G cellular data performance with a Jetpack modem.

We know in most areas that Verizon primarily operates 4G/LTE in the 700MHz band, although Verizon has been actively deploying LTE into all of its other bands too — 850MHz, 1900MHz, and 1700/2100MHz.

The Wilson 301103 magnetic-mount antenna spec sheet reveals it will provide a gain of 1.9 dBi in the 700MHz frequency range, and includes the attached RG-174 ten-foot cable and connector in this spec.

That is not much gain in the 700 MHz band, and if we add an additional 10' RG-58 extension cable, we attenuate the signal an additional 1.3dB, plus 0.5dB for the connector. This leaves us with 1.8dB line loss — the gain from the antenna is negated by the extension!

But let's look at this same antenna in the 850/1900MHz band that Verizon uses primarily for voice and 3G data. At 1900MHz, the antennas has a gain of 6.12 dBi, and at 850MHz a gain of 5.12 dBi, so we know that for voice calls this antenna will work well, even with an extension cable.

This antenna was clearly designed as a dual-band 850/1900MHz antenna, and only marginally performs in the 700MHz bands.

In other words, this is a bad antenna choice for LTE and mobile data!

What would be a better antenna in this situation? Next, let's look at the popular Wilson Trucker antenna, model 311133. In the 700MHz band its gain is 3.1dBi; at 850MHz, 4.1dBi; and at 1900MHz, 5.1dBi.

So far the Trucker looks promising — but at the 1700MHz/2100MHz AWS frequencies used by Verizon's XLTE service, this antenna actually has a negative -1.8 and -2.1 dBi gain — turning it into a bad choice for anyone with a modern XLTE-capable device.

Adding an amplifier to the mix can improve overall gain quite a bit, even just using the bundled antenna. For example, let's look at the Wilson Sleek 4G 460107 cradle-style booster (now known as the weBoost Drive 4G-S).

Again, focused on boosting 4G data, we insert the Jetpack hotspot into the cradle and use the provided stubby omnidirectional antenna (model 301126) magnetically mounted on a good ground plane on the roof.

This antenna supports all the Verizon LTE bands, but it only has a 2 dB gain in the 700MHz band, so it initially seems not very powerful. But at 1900MHz it has a 5.1 dB gain rivaling the much larger trucker, and in the 1700Mhz/2100MHz AWS bands – instead of a negative gain the little stubby provides a solid 4.5 dB / 5.6 dB gain!

The amplifier then provides a max 23 dB of additional gain on top of what the antenna alone provides.

So even with a low-performance antenna, the booster should outperform the Trucker antenna alone in almost every circumstance.

And if you plug a higher gain antenna into the Sleek, you will further maximize your performance potential.

From the above examples, you can see that performance is dependent on a variety of factors that interrelate. Varying only one factor may not provide the maximum available performance in a given situation.

You must take all factors into account when designing your system.

Measuring Antenna Performance

Getting a valid measurements to evaluate antenna performance can be tricky without specialized gear.

There are many variables that affect performance, and some of them are out of your control. But the bottom line is:

For Cellular – Are your voice calls going through without issue? What is your data speed as tested with a tool like speedtest.net, or speedof.me? The actual results are what count more than anything else.

For Wi-Fi – As above, what is your data speed using the speed testing tools? Over what range are you able to remain connected?

While bottom-line speed results are what count, you can also view the raw connection strength to your cellular tower or to your Wi-Fi access point with most phones, modems, and routers.

They all vary in how they provide this measure. What you are looking for is something more granular than just signal bars — look for the signal RSSI or the raw measurement of received signal power in dBm.

If you change antennas, move antennas, add extension cables, or add amplification, comparing before and after measurements of these numbers can tell if your signal improves.

If it does improve, you would typically have better quality voice or faster data speeds.

On the Roof – Mounting Methods & Examples

We know that for both Wi-Fi and cellular signals the main culprits that affect your results are:

- Poor line of sight (LOS) to the source signal on the cellular tower or Wi-Fi access point.

- Radio equipment that is not powerful enough, including poor antenna choices.

Mounting your high-quality antenna — for either Wi-Fi or cellular use — on the roof of your RV or vehicle helps to overcome poor LOS (line of sight).

Getting the antenna above obstructions such as other RVs will greatly improve your overall performance. Just doing this often improves the signal enough so that amplifiers or other signal-boosting techniques are not required.

Let's look at some common mounting methods for both cellular and Wi-Fi antennas.

Interior Mounts

Though the altitude that comes from roof mounting is ideal, sometimes it is not practical. In those cases — with a little trial and error at each new stop you can find an optimal interior window to place your antenna inside of.

This photo is of a cellular panel

antenna with a temporary mount placed on a motor home dash. You can see a Wilson cellular amp sitting next to it.

Although performance would be improved by mounting this panel antenna above the roofline of the motor home, this provides sufficient gain in many circumstances. The panel antenna does need to be pointed toward the cellular tower to help, though.

Side Mounting

Permanent or temporary side mounting of larger antennas on vehicles works well. Especially on RVs, side mounting on the driver's side offers some protection from tree limbs and low bridges that might affect an antenna mounted directly on the roof.

This picture shows a permanently mounted Wilson Trucker cellular antenna with a flexible spring base mounted next to a temporarily mounted Wi-Fi antenna that is a replacement for a Wi-Fi router's built in "rubber ducky" antenna.

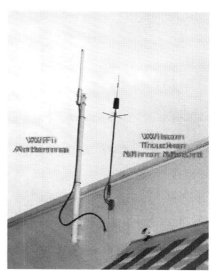
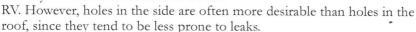
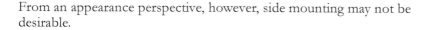

On this RV, the Wi-Fi antenna is dismounted and snapped into a carrying bracket for travel – you can see the bracket just above the window awning. The antenna mast is made from a piece of PVC tubing and screwed to the side of the RV.

The obvious downside to this method is that you have holes in the side of the RV. However, holes in the side are often more desirable than holes in the roof, since they tend to be less prone to leaks.

From an appearance perspective, however, side mounting may not be desirable.

Roof Mounting

The key challenge with roof mounted antennas is protecting them from damage while underway.

This is commonly accomplished by using shorter height antennas, or taller antennas that can be retracted flat.

Here is a Wilson low-profile cellular antenna attached to the side of a solar panel mount with a piece of aluminum plate.

This mounting location allows the antenna to function while traveling, but has the disadvantage of potential shading of the panel. This is not a magnetic-mount antenna, but is an NMO (New Motorola) style.

This pictures shows a Wilson Trucker cellular antenna permanently mounted to a batwing TV antenna. It is simply attached to a stainless steel handle screwed to the batwing base. The normal mirror mount is used.

The disadvantage of this method is that the antenna is not available for use when driving because the antenna is lowered. The advantage of this method is that the antenna is out of the way on the roof and is up high when the batwing is deployed.

It does not interfere with normal use of the TV antenna.

Life Without a Batwing

With most RVs no longer using batwing TV antennas, a retractable batwing is often no longer a convenient mounting option that allows taller antennas to lay flat while underway.

Direct roof mounting of larger antennas and CPEs is now required.

In most cases I like to use magnets for this. Use of magnets to attach the device to the roof allows for a tree strike to "flip" the antenna or CPE over – hopefully without damage. It is then a simple matter to right it.

I typically use the small, round, fifteen pound magnets you can find at the home stores. Look in the hardware section.

Attach a ferrous metal plate to the roof using adhesive caulk – just like for a ground plane. That gives the magnets something to "stick" to. Then use any ferrous metal to attach to the antenna or CPE.

Shown below is a Ubiquiti Bullet attached to a metal water filter bracket for flat roof attachment. A right angle adaptor attaches a stubby Wi-Fi antenna to the Bullet. Magnets are simply placed on the metal plate and the bracket laid on top of it. You only need three magnets – 45 lbs of attraction is more than enough.

Orient the device cross-ways on the roof; that keeps water from being forced into joints, and more importantly, allows the device to roll "sideways" if struck by a tree limb.

This next photo is a WirEng BoatAnt mounted to a circular metal plate on a roof.

The shaft that it is attached to is plastic pipe nipple screwed into a metal floor flange. This flange is simply placed on the magnets with the shaft facing forward. The shaft offers some protection to the antenna, while the plastic will not interfere with the signal. Do not use a metal pipe nipple, since it may affect the signal.

Motor Mounts

To mount a taller antenna flat on the roof – when a batwing is not available – look into motor drive mounts.

 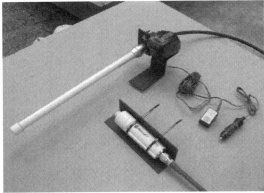

Pictured above is a motor-drive mount that lifts an 8dBi Wi-Fi omni antenna used with a Ubiquiti Bullet as a CPE. The Bullet is mounted separately, flat on the roof. The motor drive is 12-volt and is used to raise and lower the antenna much like the TV batwing does. There is a switch to control it mounted in the interior of the vehicle.

A motor mount can be mounted on a vehicle or on an RV roof. It is attached to the roof with VHB tape, or 3M 5300 adhesive. Or, you can use the magnetic mounting system mentioned earlier as well.

Roof mounted antennas should be kept to the left side of the roof, if possible – lessening the risk of being hit by a low branch on the curb side of a rig.

Mast / Flagpole Mounting

Mast mounted antennas have the advantage of height, but the disadvantage of setup time. They often work best when used to compliment a roof mounted antenna only when needed, not as a primary antenna.

In these photos we see two temporary mounts using a 16' collapsible painter's pole repurposed as a mast. This pole reduces down to less than 8' so it stores fine in most RVs with coach width storage space. There are smaller poles available as well. The picture on the left is a Wilson Trucker mounted to the top of the pole with electrical tie-wraps.

The antenna wire goes through a window in the slide or is shoved around the slide seal. This method gets the antenna quite high. In this particular location, the cell tower was 18 miles away, and there was no detectible signal at the RV. With use of the antenna alone, there was a barely detectible signal. The addition of an amplifier gave solid voice and 3G data.

The painter's pole is attached to the ladder with tie wraps. Some people permanently attach the painter's pole to the ladder and simply lower the top section when moving locations. The extra cable is looped around the ladder for travel.

The second picture is a Wi-Fi CPE attached to the painter's pole with a clamp mount. The top part is an 8dBi omni antenna, and the bottom part is a Ubiquiti Bullet.

Again, this is a temporary mount and is attached to the ladder in the same fashion. The device is powered by Power Over Ethernet (POE) and is used to improve long-range Wi-Fi capture.

Another way to gain some altitude is via a collapsable flagpole – providing a way to both fly your colors and enhance your signal. There are many RV flagpole kits available – the FlagPole Buddy mount (www.flagpolebuddy.com) is particularly impressive and provides an easy way to quickly raise and lower a flagpole.

The FlagPole Buddy comes in sizes ranging from 12' to 22' high.

It is an easy thing to strap a Wi-Fi CPE like the Ubiquiti NanoStation or a WiFiRanger Elite to the top of a flagpole, getting above all nearby obstructions.

One note on using a flagpole: you want to make sure that you do not attenuate the signal too much with extra cable length, which could easily negate any benefit from placing the antenna higher up.

Because of this, in most cases very high mast mounting is best used with equipment that uses Ethernet to carry the signal – which means Wi-Fi capture and not cellular.

Using a cellular antenna at the top of a tall mast requires careful planning and a longer low loss cable than just roof mounting requires.

If you feel a need to get even more altitude than a flagpole can provide, a retractable mast from a TV news truck can get your antennas over 60' into the air – if you really want to get crazy.

Directional Antennas & Aiming

Mounting directional antennas requires the ability to point them towards the signal source.

Shown above is a CPE with a directional panel antenna used to capture Wi-Fi. This is permanently mounted to the RV ladder with PVC schedule 40 pipe. The pipe structure is built with a 1.5" pipe inside a 2" pipe. The inner tube is longer than the outer tube and is used to extend the CPE above the roofline when not traveling. The inner pipe has 8 holes in it and is rotated to point towards the signal source. The outer sleeve has two holes. The inner holes are aligned to the outer holes and pinned in place to lock in a given direction.

For travel, the inner pipe is lowered so that the CPE is below the top of the ladder, and it is pinned in place securely.

The required Ethernet cable is routed to a jack on the side of the RV. This mounting method can be used for any directional antenna.

A directional antenna can also be mounted to the batwing antenna. Here we see two versions of a WiFiRanger Mobile unit mounted to the TV batwing. Although these both use an omnidirectional antenna (a Laird stubby and an 8dBi omni), you could also mount a directional Ubiquiti NanoStation in the identical fashion.

The NanoStation uses a directional antenna and performs at a far higher level than the mobile unit using the Laird antenna.

The con of mounting a directional antenna using the batwing is that the TV cannot use the batwing for aiming at the same time, since the antenna is dedicated to pointing at the Wi-Fi or cellular source.

Pictured below is a directional CPE simply screwed to the top of a batwing. This is a very simple but effective mounting technique. Power is supplied over Ethernet, and there are no issues with cable length as there are with analog antennas.

Getting Creative

You can fabricate antenna mounts from most anything to creatively solve mounting challenges.

PVC works well in a lot of cases and has the added benefits of not rusting, being readily available, and easy to work with.

For example – the WirEng BoatAnt cellular antenna shown here is intended to be mounted to a pole, and has an integrated mounting plate. But it can be roof mounted by adapting some schedule 40 PVC piping and using caulk – as shown.

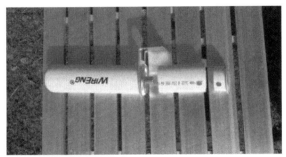 

Cable Routing & Management

The biggest issue most people have when mounting equipment to their RV is the fear of penetrating their RV roof. There are various ways to do this with little issue.

The best way, if it is possible, is to drill a fairly large hole in the roof and run a conduit from the communications cabinet directly to the roof. Then use caulk to mount an electrical utility junction box over the conduit. This provides an easy way to route cables as required – and the conduit makes it easy to change or add cables later.

Even if not using conduit to enter the RV, use a junction box to cover the entire hole in the roof. This provides the most weather-resistant seal possible.

The wires exit the junction box on the roof via a weatherproof cable connector. These connectors may be found at any Home Depot or Lowe's in the electrical section. Shown here is the communications box with a

Category 5 Ethernet cable and an RG-58 cellular antenna cable exiting it. On the right is the solar combiner box.

Both of these have cables exiting the bottom of the box and entering the RV directly through the roof. For the wires exiting to the RV roof, try to face the cable exits to the rear of the RV, where possible. This helps keep water from being driven into the box by wind as you drive. If you cannot exit from the rear, then use a side exit. Avoid a front exit.

Self-leveling caulk compatible with the roof type is used to hold the box in position and simultaneously seal it. No screws are needed or used.

Wires can be held in place on the surface of the roof by pools of caulk – make sure the caulk is compatible with the roof material.

Place a puddle of caulk on the roof and lay the wire through it. Use tape on either side of the puddle of caulk to ensure the wire stays pressed into the puddle. When dry, remove the tape and place some additional caulk over the wire and the original puddle of caulk. This will ensure the wire will not pull out of the caulk. Use your best judgment about the distance between puddles – four to five feet is usually close enough.

If you have a number of wires running together, it is generally acceptable to embed two of them in the caulk puddles and then simply wire-tie the others to those two. They will not go anywhere. Make sure you use UV-resistant wire ties. Even these will degrade in a few years, so keep your eye on them when doing your roof maintenance.

Another retention method that works well is to use a small piece of Eternabond sealing tape over the wire. A two inch long piece is adequate place them every 4 feet or so. Generally, the caulk method is preferred since it is far easier to remove, if required.

Also, make sure that any wire on the roof is UV resistant. Antenna wires are fine to expose to the weather. Ethernet cable needs to be outdoor-rated cable. Do not use plenum-rated Ethernet cable if you are wiring a device yourself.

An alternative to making a new hole in your roof is to take advantage of an existing vent opening. The refrigerator roof vent is often a convenient place to run wires down from antennas.

Conduit

Any time you run a wire, think ahead. What if you ever need to upgrade or replace it?

Using cheap plastic conduits (easily found at any hardware store) to connect key locations in your RV makes future wire runs and antenna upgrades a lot easier.

You can use conduit running forwards and backwards in your RV, or even to the roof.

A 1" conduit is a good minimum size, but for more flexibility you can go larger.

Some RVs even come with conduit pre-installed in key locations.

In my case, I always install a dedicated "wire entry" box that sits on the roof and has 2" conduit that enters the coach inside it. This makes it very easy to add wiring whenever I need to.

Shown below is the box on my roof with the conduit. This actually goes to my basement, but has a cutout in my communications cabinet for antennas, ethernet wire, etc.

The picture shows the box before any wiring is added. The red wire is a pull-wire to the basement, which can be used to pull new wires through.

To add an antenna I drill a hole in the rear of the box and thread the cable down to the tech cabinet. The cable enters the box through a waterproof connector, also shown below.

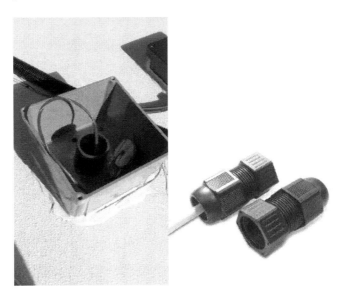

Satellite Internet

Before cellular internet and prevalent Wi-Fi hotspots became the norm, satellite internet was the ultimate option for getting online at better than dial-up speeds while mobile.

Back then, if you walked around any high-end RV resort, the signature blue glow of Datastorm dishes mounted on roof tops meant one thing: the people inside were online, while everyone else made do with clubhouse dial-up, or nothing at all.

But sadly, the glory days of relatively affordable, available-almost-everywhere satellite broadband have been fading into the past for mobile users. *But a renaissance may be at hand!*

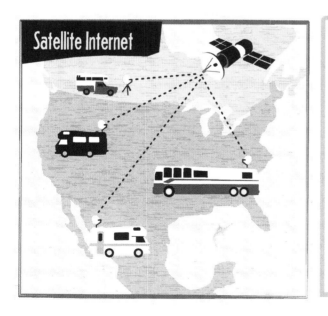

Quick Glance

Pros

Signal (almost)
Everywhere

Cons

Latency

Bandwidth Caps

Slower

Expensive/
Complicated

Satellite Internet

During the Summer of 2015, satellite internet began to get interesting again with some exciting new options debuting on the market.

And assuming certain ambitious plans ever (literally) get off the ground, there are signs of a full fledged satellite internet revolution coming in the years ahead.

Satellite is a great option for some situations.

Especially those situations where it is the ONLY option.

The Past: Glory Days Fading

A decade ago, satellite made sense.

In the days of 2G, consumer satellite offerings from HughesNet could run circles around any cellular plan in both speed and coverage.

But as 3G cellular service spread - it was almost always faster and cheaper.

Increasingly widespread cellular 3G and 4G coverage maps ate into the 'go anywhere' advantages of satellite, leaving fewer and fewer customers willing to put up with the costs and considerable headaches associated with orbital internet.

Many RVers with satellite internet systems demoted them to secondary connections, and then canceled service entirely.

With fewer customers to serve, satellite internet companies that had targeted RVers went out of business, or switched focus to more lucrative markets.

Especially in a world of 4G/LTE cellular, satellite just couldn't keep up.

The Slow Lingering Death of Classic HughesNet

HughesNet (www.hughesnet.com) has long been the leader in consumer-grade satellite internet, and in some rural areas HughesNet remains one of the few affordable options for getting online.

Though HughesNet never officially directly supported mobile users, there used to be several companies that provided the equipment and training to allow for tripod-mounted HughesNet dishes to be taken on the road.

Commercial partners also officially resold HughesNet service (at a slight markup) paired with auto-aiming roof-mounted dishes, like the classic MotoSAT Datastorm.

But as HughesNet has evolved to focus consumer offerings exclusively on next generation residential spot-beam service, it has grown increasingly eager to migrate customers away from the old satellites and service plans.

And unfortunately for RVers – the new spot-beam equipment and plans from HughesNet are NOT mobile compatible, and are only suitable for fixed location installation.

Though some grandfathered customers remain roaming the country using older HughesNet HN7000S satellite modems with legacy service, as of 2014 HughesNet has made it nearly impossible to activate new consumer-grade service using older equipment.

The days of taking a cheap HughesNet plan on the road are essentially all but over.

Starband: RIP September 30th, 2015

Though HughesNet openly tolerated many tripod-based mobile deployments, StarBand had been the only satellite internet service that officially supported customer-aimed portable tripod installations, and thus was formerly a popular choice for some RVers.

But with no forward evolution to faster spot-beam technology planned, StarBand had increasingly little to offer residential customers, and over the past few years was falling ever further behind.

StarBand last year announced that it was shutting down service entirely, turning off the network and shutting its doors on September 30th, 2015.

Exede, WildBlue & dishNET: Fixed Locations Only

HughesNet and StarBand were not the only residential satellite internet service providers.

WildBlue, Exede, and dishNET all advertise residential satellite internet plans with hefty data caps and download speeds over 10Mbps, often for well under $100/mo.

So of course there was a steady stream of RVers hoping to find a way to take these services on the road.

But these plans are sadly NOT mobile friendly.

Satellite Internet

WildBlue (www.wildblue.com) and Exede (www.exede.com) are both subsidiaries of parent company ViaSat – with WildBlue being the older generation service, and Exede the latest and fastest.

Both of these services have used spot-beam technology from the start – making them totally unsuitable for mobile usage.

And while it is certainly possible to take satellites TV service from Dish Network on the road – the widely advertised companion dishNET (www.dish.com/dishnet/) internet service is actually contracted out to Exede or HughesNet behind the scenes, meaning spot beams that eliminate the possibility of mobility.

Don't be tempted to even try signing up for these services if you want mobile flexibility. New service requires lengthy contract commitments, and your account is provisioned to only work at your fixed "home" address.

These services are only useful for RVers who need seasonal service – with a fixed dish installed at a summer or winter home, and service suspended while away.

The Present: A New Dawn

With HughesNet phasing out mobile compatible plans, and StarBand gone, RVers looking to get connected by satellite were nearly out of options.

But new in 2015 there are now once again providers of "satellite broadband" that are supportive of mobile users.

The dream of connectivity everywhere is alive!

Here are all the currently viable satellite internet options RVers should consider:

RV DataSAT 840 / Insta-Sat Service

In July 2015, commercial satellite gear provider Mobil Satellite Technologies (www.mobilsat.com) announced the RV DataSAT 840 – a **new** consumer-focused satellite internet terminal, designed specifically for RV roof mounting!

With MotoSAT long gone, this is billed as *"the only automatic satellite Internet antenna designed especially for consumers"*.

Satellite Internet

The RV DataSAT 840 is a big dish — and it takes up a substantial chunk of roof real estate — potentially casting a serious shadow on any nearby solar panels.

A full setup (including the modem and installation) currently costs over $6k. This is not cheap, but if you've been dreaming about the glory days of RV satellite internet — at last there is an option.

Dealers for the system include Oregon RV Satellite Service (www.oregonrv.net) and Montana Satellite Supply (www.montanasatellitesupply.com).

The Insta-Sat service plans that accompany the RV DataSAT 840 allows pay-as-you go data to be purchased without a contract, monthly fee, or long term commitment.

This is great for RVers who only need satellite occasionally throughout the year — letting you activate service only when you need it. You can even activate over the satellite connection — right from the middle of nowhere.

The Insta-Sat 4Mbps speeds and limited data buckets aren't very exciting by cellular standards, but as a fallback service when out of cellular range this is actually a pretty enticing offering for those who need this kind of reliability.

All Service Plans Have The Same Rated Speeds:
up to 4 Mbps down x 512 Kbps up

Plan ID	Mobile Plan	Price	Good For
Insta-Sat **ISAT5**	5 GB	$125	1 Month.
Insta-Sat **ISAT25**	25 GB	$400	3 Months.
Insta-Sat **ISAT50**	50 GB	$600	6 Months.
Insta-Sat **ISAT100**	100 GB	$1,000	12 Months.

And for those who are addicted to Netflix — there is now a $199/mo RV Entertainment plan from Insta-Sat that uses a technology called NightShift to download your preselected Netflix shows during a free data period overnight, so you can watch the shows later without burning up any data.

More details on plans and options can be found here: www.rvdatasat.com

HughesNet Spot Beam Service from RTC

Something once thought nearly impossible is becoming real:

Next generation spot-beam satellite service – with mobility support!

Traditionally – consumer-grade spot beam service has been strictly limited to fixed installations. Even if you physically were able to move and re-aim a dish, the satellites would refuse to acknowledge you at your new location.

Policies however are evolving – and HughesNet has been quietly running a trial program throughout 2015 coordinated by reseller Real Time Communications (www.rtc-vsat.com) that for the first time enables an affordable spot-beam satellite service to be provisioned to automatically change spots.

Even more surprising – you do not even need an expensive auto-aiming roof-mounted dish to do so – RTC is now selling a tripod satellite kit with everything you need to get connected for around $1000.

RTC is currently offering mobile HughesNet SpaceWay (third generation satellite) service with 20GB usage for $99/mo, with speeds of 5Mbps down and 1Mbps up.

This may seem slow by LTE cellular standards, but for satellite these speeds are very decent.

HughesNet's faster fourth generation "Jupiter" service will not have nationwide coverage until the next satellite is launched in late 2016. Once the Jupiter service is available nationwide – lower data prices, faster speeds, and additional service offerings will become available for mobile users – likely towards the end of 2016 or in early 2017.

These HughesNet plans require an 18-month contract and commitment to a monthly service plan – locking RVers into paying for satellite service even when in places they do not need it, though temporary service suspensions of up to six months per year are allowed.

For RVer focused hardware and support with these HughesNet tripod options, check out MobileInternetSatellite.com who has partnered with RTC to bring these new mobile options to the RV market.

Legacy HughesNet HN7000S Service Offerings

If you are willing to fend for yourself getting service activated, you can often still find old HughesNet Datastorm and tripod systems – occasionally literally being given away.

The key is finding systems using the old HN7000S satellite modem, a sure sign of compatibility with the still active nationwide HughesNet Ku-band service, which does not use spot beams or place any limits on mobility.

Start your hunt for used gear with the Datastorm Users Group (www.datastormusers.com), various RVing forums, and check Craigslist and eBay. Nobody wants to ship big and bulky satellite gear, but if you are willing to go and help somebody take a dormant system off their roof, you can often get satellite internet equipment for a song.

If you are looking to buy older gear, here are some resources:

- OregonRV.net buys up old equipment, and refurbishes and resells it to consumers.

- RFMogul.com is a group of former MotoSAT engineers who provide new Datastorm controller software and upgrade options for old controllers too.

The real challenge with legacy HughesNet service isn't getting hardware – it is getting service activated.

As of mid-2014, HughesNet has made it nearly impossible to activate new service on HN7000S modems, so expect a challenge unless you are taking over an active grandfathered account.

The best way around this is to set up legacy Ku-band service is through a commercial reseller.

These providers are able to still support legacy HughesNet HN7000S service – but the required "commercial grade" plans to activate an HN7000S can be painfully pricey:

- Montana Satellite (www.montanasatellite.com) – The current cheapest HN7000S service plan is $99/mo for 1Mbps download speed and 350MB per day usage.

- Mobil Satellite (www.mobilsat.com) – Now focused on launching the new RV DataSAT 840 and Insta-Sat service, but they can still activate legacy HughesNet hardware.

BGAN (Broadband Global Area Network)

If you are more interested in basic connectivity rather than broadband speeds and bulk transfers, Inmarsat's BGAN network is a potentially interesting option.

Rather than a big bulky dish and separate modem, a complete BGAN satellite system is about the size of a laptop computer – and it does not need to be carefully aimed and can usually be set up in a matter of minutes.

And BGAN service covers the entire planet, other than the poles, perfect for sailing nomads or expeditions who plan to roam continents.

But…BGAN is slow and extremely expensive!

A basic BGAN terminal like the IsatHub iSavi (www.groundcontrol.com/IsatHub.htm) costs around $1,300 – and is capable of "broadband" speeds of just 384Kbps down and 240Kbps up.

Faster, fancier, and more rugged BGAN terminals go way up in cost – but even the highest end plans and radios top out at less than 512Kbps.

Download speeds of 384Kbps are not enough to even dream of streaming YouTube, but it is plenty for checking email and some basic old-school super-slow web surfing.

But it is the data costs that will kill you: starting at $3.85/MB ($3,850/GB!) sending data via BGAN can quickly get ridiculously expensive.

BGAN terminals can also double as satellite voice phones as well, making calls anywhere in the world for $0.99/minute.

The key to using BGAN is extreme data-usage management – focusing on email and other text-centric communications rather than interactive graphical web surfing.

Low Earth Orbit Options

Most satellite options rely on carefully aiming at satellites in fixed locations in the sky – parked in geostationary orbit over the equator 22,236+ miles away.

Satellite Internet

It takes a carefully aimed, traditional-looking dish to communicate over those extreme distances – or an expensive and slow BGAN terminal.

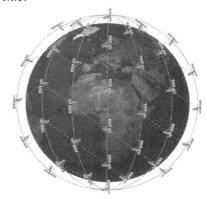

Satellites in lower orbits can be easier to talk to, but they are always in motion through the sky and any given satellite will quickly pass over the horizon and out of sight.

But a constellation of satellites working together can provide constant coverage from much lower altitudes.

There are two current low earth orbit satellite constellations worth considering:

Globalstar

Globalstar (www.globalstar.com) has a constellation of 40+ satellites 880 miles up, and has long been a provider of satellite telephone service.

Now Globalstar is offering a $999 box called Sat-Fi, which they like to describe as a satellite-powered Wi-Fi personal hotspot that supports unlimited data with plans currently starting at just $39/mo!

Before you rush to sign up for this perfect-sounding solution…

The catch with SatFi is that the data speed is 9.6Kbps (vintage mid-1990s modem speeds), and the included data only covers basic email, SMS text messages, posting to FaceBook and tweeting on Twitter (not reading), and very little else. The $39/mo plan also includes 40 minutes of satellite phone voice minutes – which can also be used to make very slow general-use data calls.

For $149/mo – you actually get unlimited voice and data calls.

The data speeds are so slow that you would never be able to interactively surf the modern web. But it you have extremely low bandwidth needs or are primarily voice-centric – this is actually a very affordable option.

Iridium

The Iridium (www.iridium.com) constellation provides global coverage with 66 swarming satellites orbiting 475 miles up.

Iridium is known for powering handheld satellite phones that enable voice calls anywhere on the planet, but the network does support limited data service too.

Like Globalstar, Iridium launched a satellite Wi-Fi hotspot in 2014 – the Iridium Go! – also priced at $999. Plans currently start at $49/mo, or you can get unlimited (but extremely slow 2.4Kbps) data for $124/mo.

At that speed, uploading a single high-resolution image might take hours, and even a basic web page might take 5–10 minutes to display. This is not a device for surfing on!

The other big gotcha is that all data usage must go through Iridium's special apps on your phone or tablet – you cannot get a laptop online or use arbitrary apps.

DeLorme inReach

Another more consumer-friendly option that uses the Iridium network behind the scenes is the DeLorme inReach (www.delorme.com).

These rugged handheld devices cost $299 (or $399 with navigation features) and provide continuous remote tracking and support for 160-character, two-way text messages that work essentially anywhere on the planet – with plans ranging between $12 and $99/mo.

If you are headed out on a remote expedition, this would be a great bit of technology to have with you.

SPOT Tracking

Very often when you are out in the boonies far from other communication options, your communication needs are actually very simple.

You either want to let your loved ones know "I'm fine, and this is where to find me..." or you want to let everyone know "Send help – here is where I am!"

For these basic needs, a big expensive satellite system is overkill.

The basic SPOT Satellite Messenger (www.findmespot.com) device combines a GPS with a one-way satellite transmitter (powered by Globalstar) that works almost everywhere in the world (there are a few gaps in coverage, but not many).

Just press a button to "Check In" or another button to "Send Help."

SPOT service currently costs $99/year, and automatic tracking that logs your location to a shared Google map every 10 minutes costs $49/year.

The big downside of SPOT is that it is broadcast only – there is no way for you to receive a message.

But it is affordable, and the safety and increased peace of mind can be very worth it for many travelers in particularly remote areas.

The Future: Revolution Coming?

The present state of satellite internet as of early 2016 still leaves a lot to be desired. But there are some exciting future prospects, just over the horizon.

In geosynchronous orbit 23,000 miles up things are going to be getting much faster:

> **ViaSat:** ViaSat's next satellite, ViaSat-2, has been delayed until early 2017 – but when it launches it will offer 350Gbps of total capacity, more than doubling what ViaSat-1 can provide.
>
> This opens the doors to faster and cheaper Exede service for fixed installations, but the network is being designed to make it easier to enable mobile installations too – so hopefully ViaSat will soon partner with a company to bring service to the RV market.
>
> ViaSat has also shared plans for providing backhaul service to remote campgrounds and RV parks, potentially speeding up campground Wi-Fi in places where there had formerly been no better options.
>
> Further out, ViaSat-3 is on the horizon with a jaw-dropping 1,000Gbps capacity (more raw data capacity than ALL commercial satellites currently in orbit – combined!) – but the first of the three planned ViaSat-3 satellites is not scheduled to launch until late 2019 or early 2020, meaning that we have a long time to wait before one-terabit service is parked above us.

And in low earth orbit, there are three new swarms slated to come online:

> **SpaceX:** As of early 2015 – Elon Musk's SpaceX (www.spacex.com) has launched a new initiative to develop a "global communications system that be would be larger than anything that has been talked about to date" – with Google as a key investor.

Satellite Internet

OneWeb: Not to be outdone, billionaire Richard Branson and Virgin Galactic are backing another new satellite internet startup, OneWeb (www.oneweb.world), that is "creating the world's largest ever satellite constellation."

OneWeb is even further along than SpaceX, and has contracted with Airbus to start manufacturing 900 satellites with launches slated to begin in 2018.

Both SpaceX and OneWeb are aiming to put hundreds of very advanced satellites into a low earth orbit swarm, low enough to avoid the latency issues of higher altitude satellites.

Iridium Next: Slightly higher up the orbital ladder – the old Iridium communications satellite swarm (best known for satellite voice phones) is slated to be replaced with 66 new "Iridium Next" satellites with substantially upgraded capabilities. The first launches are already underway, with the upgrade targeted for completion by 2017.

All these swarm-style satellite constellations have orbital paths synchronized, so that at least one satellite from each service will always be in sight and passing overhead no matter where you are on Earth.

And thanks to the lower altitude, precise dish aiming will not be required.

It is a lot easier to communicate with a satellite 500 miles away, as opposed to beaming a signal at traditional geosynchronous communication satellites locked in a fixed location in the sky 22,236 miles above the Earth's equator.

Drone-Powered Internet: Taking things to an even lower altitude – Facebook and Google have both actually been developing and test flying giant high-altitude drone aircraft that have the potential to stay aloft for months at a time, providing internet coverage to huge swathes of territory below.

Think of it like a satellite – only much lower.

And Google has even been testing stratospheric balloons to act as Internet relays with the project tagline: "Balloon-Powered Internet For Everyone"

All of these projects and companies have the potential to revolutionize satellite broadband... but not overnight.

The earliest any of these next generation networks will be live is 2018, and even the most basic pricing and service details will likely not be publicly revealed anytime soon.

Realities: Today & Tomorrow

As fun as it is to fantasize about connectivity everywhere – for most people satellite internet as it exists today just doesn't make a lot of sense, and comes with too many tradeoffs.

And even once the next generation satellite systems are fully deployed – they will at best still be complimentary systems to terrestrial cellular.

> **In an urban or suburban area, the truth is that satellite will never be able to compete with fully built out LTE on the ground.**

But on the other hand – in many remote areas it will never make sense to fully build a network of cell towers, and no matter how much cellular companies expand, there will ALWAYS be gaps in coverage.

The ideal connectivity future involves a mix of satellite and cellular, with service roaming seamlessly to the best connection possible wherever you happen to be.

It will take a while for all the essential pieces to fall into place, and for the necessary partnerships and technologies to emerge.

While we wait – we can look to the skies and dream!

Concepts & Frequently Asked Questions

Satellite internet can be confusing. It is important to make sure that you understand the basics.

Here are some of the most frequently asked questions and confusing topics explained:

Satellite TV and Internet Are Not the Same!

Receiving a signal from space isn't particularly hard.

Transmitting a signal back to a satellite on the other hand is where it gets tricky.

Satellite TV dishes are receive-only devices, and have no capability to transmit. Internet usage however requires two-way communication – and thus much larger and more complicated gear on the ground.

Satellite Internet

Some people get confused because they see satellite TV providers Dish Network and DirectTV (now owned by AT&T) advertising bundled packages that include internet service – but this is typically not satellite-provided internet.

These bundled plans are intended for stationary satellite TV consumers to combine their TV, internet, and phone bills into one. The satellite TV provider contracts out to local DSL or partner cable companies to provide the actual internet service – usually relying on a hard-wired connection.

Even when they do offer actual satellite data plans (like Dish's dishNET – which is actually provided by either Exede or HughesNet behind the scenes), these are provided primarily to offer service to rural customers outside the range of cable and DSL, and they are strictly for fixed-location installs only.

In other words – not mobile friendly at all.

NOTE: Though the dishes look similar, most satellite internet systems are strictly for internet service, and are NOT compatible with any satellite TV services. If you also want satellite TV to go along with satellite internet, you'll actually need a *second* dish on your roof!

The Signal: Spot Beam vs. Broad Area

Traditionally, communication satellites broadcast a signal on Ku-band microwaves that could reach an entire continent.

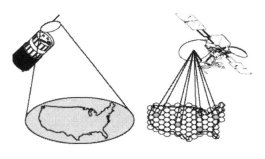

This is great for mobile users – the satellite doesn't know or care whether you are in Boise or Boston, in the Black Rock desert or back-country Georgia.

But it is also horribly inefficient – with every user assigned to a particular satellite and channel having to share the signal.

Newer satellite services such as WildBlue and Exede and HughesNet Gen3 and Gen4 use **spot-beam technology** to cover the nation with many small focused signals using higher frequency Ka-band microwaves rather than a single blanket Ku-band signal – allowing for many more users to communicate at once.

This allows for cheaper service and faster rates. But it also means that if you travel more than about 100 miles from your "home" address, your satellite service will no longer work at all – assuming you can even physically move and then aim the dish correctly, since spot beam receivers are especially sensitive to needing absolute precise aim.

Why Not Just Change Spots?

It is technically possible for a satellite receiver to "change spots", but up until very recently none of the consumer satellite systems provided any support for automatically or even manually changing your assigned beam.

Relocating a dish to a new location thus usually required a certified installer aiming the dish while on the phone with the network operations center coordinating the move, and even with professional help at most you might be allowed to relocate service only once a year.

Because of the mobility-hostile nature of spot-beam service, RVers using satellite internet have been limited to older Ku-band options that are slower and more expensive.

Worse – the leading consumer satellite internet provider, HughesNet (www.hughesnet.com), has been actively phasing out their legacy Ku-band service, making it almost impossible to activate older equipment on a consumer-priced service plan.

Policies however are evolving – and HughesNet is beginning a trial program (covered earlier in this chapter) to enable mobile Ka-band deployments that can change spots without needing an installer's help.

This is a revolutionary shift, and once again opens the door to truly mobile satellite internet installations.

Even Satellites Have Coverage Maps!

The great advantage of satellite Internet service is that you can usually connect anywhere you have an unobstructed view of the southern sky.

But there are actually still coverage map issues when it comes to satellite.

HughesNet, for example, offers service on a dozen different satellites, each with a different broadcast footprint and varying signal strengths across the nation. Not every satellite actually serves every corner of the country, particularly if you want to travel to Alaska, throughout Canada, and down into Mexico.

Spot beam services tend to be especially localized – with spots typically not aimed far beyond the national borders of the service's target market.

Changing satellites you are assigned to is possible, but cumbersome. It is better to do some research in advance and try to get assigned to a satellite and service with coverage in the places you are most likely to go.

Tripods vs Roof Mounted Dishes

There are two ways to aim a satellite dish — via robot, or by hand.

A roof-mounted robotic system is undeniably convenient — with all the hard work handled automatically with the touch of a button.

A much cheaper alternative to a robotic roof dish is to use a manually aimed dish mounted on a surveyor's tripod anchored to the ground.

Getting a satellite internet dish manually aimed however is vastly harder than setting up satellite TV, and it can often take an hour or more to get a signal dialed in by hand.

Precise aim is extremely important since a mis-aimed dish can cause interference with adjacent users on a satellite. Non-professionals moving and aiming their own dishes falls into a fuzzy gray area for some satellite service providers, and is outright banned by others.

The great advantage of a tripod system is that it can be set up to avoid trees and other obstructions while you park your RV in the shade, an important consideration since even a single small branch can block a data signal.

An auto-aiming roof-mounted satellite internet dish however is practical to deploy even during the briefest of stops, with no setup time required.

Nothing beats the convenience of pressing a button and getting online a few minutes later, letting a robot do the difficult work of aiming at a target 22,000 miles away.

Latency & Satellite Communications

When you are using most satellite internet services, keep in mind that you are relaying everything via a satellite parked 22,236 miles above the equator.

That is a long way away, and at that distance the speed of light starts to impact everything in ways that feel completely foreign to those used to terrestrial connections.

Even with a fast broadband connection, there will always be a noticeable pause with every click as you wait for your request to make the trip from your dish, to the satellite in orbit, back down to the control station, then over the internet to the web server you are talking to, and then finally a full round trip back to you.

This delay is called latency, and it is especially noticeable if you try to have a video or audio conversation with someone over satellite.

On a high latency connection, it can actually help to say "over" to avoid accidentally talking on top of each other!

Unlimited Data Options

Worrying about megabytes, gigabytes, overage charges, and connection speed throttling can become a constant source of stress that those used to unlimited data can hardly begin to comprehend.

With limited connections, a lot of the data-intensive tasks that most users take for granted become much more complicated. Streaming video, homeschooling and remote education, transferring large files, online gaming, and even keeping your computers updated with the latest operating system and software updates all put your usage limits at risk - potentially opening the door to expensive overage charges.

Yet we live in a very cloud centric world these days – and apps are growing increasingly data hungry. Even if you work hard at it, remaining on a data diet can be an exercise in frustration.

It is little wonder that so many connected RVers fantasize about finding "unlimited" connections.

Unfortunately – truly unlimited, un-congested and fast mobile internet connections you can share amongst all your devices are hard to find.

But they're not impossible!

Many of these options are covered in more depth throughout the book, but this chapter provides an overview of the unlimited options in one place.

Cellular Unlimited Data Plans

Radio spectrum (the number of channels available to broadcast on) is a limited commodity, and only a certain amount of data traffic is physically possible on any given cell tower.

And yet, the appetite for consuming content online seems to be insatiable.

To try and keep usage under control and the profits flowing, cellular networks have policies in place to try and limit the heaviest users.

T-Mobile and Sprint currently treat every plan as "unlimited" – but plans come with a limited amount of "high speed data" that can be used before the connection is throttled to a near halt for the remainder of the month. But at least there is no risk of overage charges.

Many of T-Mobile's postpaid plans also include unlimited "Binge On" video streaming at lower resolution from partnered video services, and unlimited music streaming too.

All of the carriers except Verizon have options for new customers to sign up for unlimited smartphone data plans. But these plans are for data used on the small screen of a smartphone only, and limits come into play when sharing data with any other devices.

AT&T, T-Mobile, and Sprint also all have network optimization practices in place to keep these unlimited plans in line, meaning that after using somewhere between 20-25 GB a month on an unlimited plan, the carriers reserve the right to "deprioritize" your traffic for the remainder of the month, putting you in a relative slow lane compared to other customers – but only when connected to congested cell towers.

The carriers also each have limits on how much data from these unlimited plans can be shared with other devices – AT&T offers no support for sharing, Sprint only allows 3GB, and T-Mobile offers 14GB.

If you want high speed data beyond those limits for sharing with your tablets or laptops, you are required to get a more traditional plan with a limited data bucket. These plans are subject to speed throttling or overage charges if you go over your monthly cap.

Officially supported truly unlimited plans without all these limitations are extremely rare.

But they can be found – if you know what to look for.

The Ultimate: Verizon Grandfathered Unlimited Plans

Verizon used to offer unlimited smartphone data plans up until 2011.

While new customers can not obtain these plans directly from Verizon anymore, these old plans are still grandfathered in for existing account holders in good standing who have not given them up.

And it is actually possible to purchase and take over one of these old plans.

Unlimited Data Options

Thanks to the open access rules attached to the spectrum Verizon purchased to build its LTE network, Verizon so far has not been able to place artificial limits on how unlimited plan customers use their accounts.

This means that Verizon allows for unlimited phone plan SIM cards to be swapped into hotspots and routers – making a single unlimited line suitable for connecting the tech of an entire household.

Verizon also does not currently impose any sort of network throttling to slow down speeds after a certain usage limit is reached each month. All traffic is treated equally – even on congested towers.

These traits make Verizon unlimited data plans the holy grail of technomad connectivity.

In November 2014 Verizon made a policy change to no longer allow these plans to transfer ownership through a process known as an Assumption of Liability (AOL) without losing the unlimited data in the process.

However as of February 2016 when this book was updated, the enforcement of this policy has proven to be somewhat lax – and many of our readers have managed to recently acquire Verizon unlimited plans.

The process to acquire a Verizon unlimited plan can be daunting however, and requires careful research. The upfront costs can be substantial (lately ranging between $400 – $2000) – but the ongoing monthly fees can end up being just $45 – $95/mo for an unlimited line if you go this route.

Considering a 50GB data plan with Verizon's current plan options start at $375/month, the savings add up quickly.

There are also options for renting or leasing an unlimited Verizon line from third parties who have built up an inventory of corporate lines (who have had access to unlimited lines for a while) – with prices lately hovering around $120-$200/mo with a smaller upfront start-up fee. This route isn't officially authorized by Verizon, so there are risks involved.

Supply & demand can impact pricing – which swings considerably.

For anyone thinking they might need more than about 20GB of cellular data a month – acquiring a Verizon unlimited data plan is a very worthwhile option to consider.

Verizon Unlimited Data Plan Guide

Over the past year we have witnessed and tracked so many changes related to Verizon UDPs that more details on the process are not suitable for printing as part a book. It would be outdated as soon as we clicked 'Save'.

We keep a constantly updated and extensive guide to Navigating Verizon Unlimited Data Plans (www.rvmobileinternet.com/verizon-unlimited) in the premium members area of the online resource center.

The guide includes all of the current information we have on this process (including the latest information confirming if it's still even possible), step-by-step instructions for shopping for and obtaining the plans, vetted vendors (some who offer discounts), tips on keeping your unlimited data protected, and recent feedback from members who have completed the process.

We of course also assist in our member forums, and send out e-mail alerts if we hear of changes coming that our members need to be aware of, and we help them navigate any impacts the changes might have.

Due to the extensive nature of this guide (it's almost book length itself at 13,000+ words), and how much effort goes into keeping it updated (sometimes daily!) – it is part of our Mobile Internet Aficionados membership benefits.

> **Be aware that there is always a risk that Verizon will find a way to stop honoring these old plans at some point in the future. But Verizon actually raised the monthly price of these plans by $20 in November 2015, an unlikely move if Verizon was considering unilaterally cancelling them.**
>
> **But – do weigh the potential risks into the cost you're willing to pay to take over such a plan!**

AT&T Grandfathered Unlimited Data Plans

AT&T also has some rare grandfathered plans that some people have gone out of their way to acquire.

iPad Unlimited Plan: For a brief period in 2010 right after the very first iPad came out, AT&T offered an unlimited iPad data plan for just $29.99 per month.

AT&T quickly realized how much data an iPad can consume, especially once Netflix released an app, and quickly discontinued the plan.

But fortunately AT&T grandfathered in existing plans for those who kept them active. And the plan has been transferrable to newer iPads, including LTE-enabled devices, ever since.

Unfortunately, this plan does not include personal hotspot or tethering support – however many have chosen to either jailbreak their iPad, or transplant the SIM into a mobile hotspot to enable sharing.

Though since this is specifically an iPad-only plan, there is always the risk of AT&T seeing this usage as grounds to cancel the plan.

The plan has thus far been completely unthrottled no matter how much data you use in a month, and since the iPad can use an HDMI cable to connect into a TV set – this makes it an ideal media-streaming device for Netflix, Hulu, YouTube, HBO Go, and various TV network-specific apps. It can also be used for FaceTime, Skype, Google Hangouts, webinars, and live streamed video chats.

Prices on eBay to take over one of these plans (not including an iPad to put it in!) range from $1000 to over $2,500 at times. They are quite rare.

iPhone Unlimited Plan: Unlike the unthrottled iPad plan, AT&T's grandfathered unlimited iPhone plans are subject to network optimization after 22GB of usage a month – which can mean slower speeds while on congested towers.

These plans also do not include hotspot/tethering to share the data with other devices - substantially limiting their usefulness.

Data Only Plan: Shopping on eBay, you can sometimes find data only AT&T options. Sometimes they are old iPad plans, and sometimes they are special business lines being resold to consumers. We've not heard of any gotchas on these plans so far, but they tend to be expensive to acquire.

Reseller / MVNO Unlimited Data Plans

Occasionally a reseller or private-branded MVNO brings an unlimited data plan to market that is suitable for keeping a technomadic household online.

These tend to come and go, and we're constantly monitoring the options.

At time of this writing, the most appealing unlimited data reseller option we know of is **Unlimitedville** (unlimitedville.com).

This small company is focused on serving rural customers who don't have other options for unlimited broadband connectivity, though some RVers have taken advantage of the offerings lately.

Right now, Unlimitedville has three options, but they change frequently:

- Low-cost Sprint based data-only hotspot plan for qualified businesses, with a 2-year contract. It is completely unthrottled and has no caps. We've been told the plan is a limited time promotion that could end soon.

- T-Mobile based data only rental option which basically gets you access to a line Unlimitedville manages. It's available on a month-to-month basis, and anyone can sign up. It's unthrottled, no caps, and even works in Mexico & Canada.

- We occasionally see Unlimitedville offering a Verizon based option, which is a lease/rental. Their pricing has historically been on the higher side of the options you can find via other sources we track in our Verizon Unlimited Data Guide.

Maximizing Unlimited Wi-Fi

A lot of public places now offer free Wi-Fi access – giving anyone within range passing through a bit of connectivity, hopefully in return luring customers to patronize their business. You'll find public Wi-Fi hotspots in campgrounds, cafes, breweries, hardware stores (Lowes), libraries, motels, airports, malls, town plazas, and more.

Most offer unlimited data, and even the public Wi-Fi networks that do place limits on usage usually do so by time (purchasing a cup of coffee includes an hour of Wi-Fi access, for example) rather than by the amount of data you have used.

That's the good news.

The bad news is that a lot of public Wi-Fi networks are so slow and congested that they may hardly even be worth trying. Be sure to run some speed tests before starting any big downloads and committing yourself to nursing a pitcher of beer for hours.

Even better is making friends on the road who have both driveway parking for your RV and Wi-Fi access. Often the fastest Wi-Fi you will find on the road will be when driveway surfing, sharing a un-throttled and unlimited home cable or DSL connection.

Refer back to the Wi-Fi Hotspots chapter for more information on utilizing Wi-Fi in your travels.

TIP: Queue up your big downloads and start them before bed – they'll cause less congestion for others on the same Wi-Fi network that way, and will likely be done by the time you wake up in the morning.

Unlimited Satellite Data

Many consumer satellite broadband plans treat data usage late at night different - either offering extra "Bonus Bytes" or an unmetered "Free Zone" that avoids eating into your daytime data limits. Make sure that you understand how these rules apply to any satellite plan that you are considering - and know what the applicable special hours are in your timezone.

This overnight period is the perfect time to schedule large downloads and software updates, even if it means staying up late to get started.

> **TIP:** You can use a download manager to set a schedule, and to make sure that big tasks are cut off if they don't finish before the unlimited period runs out. Otherwise, you risk waking up to an unfinished slow download devouring your monthly daytime data allotment!

The RV Entertainment plan from Insta-Sat (www.rvdatasat.com) actually uses a special router equipped with a technology called NightShift to automatically download your preselected Netflix shows during a free data period overnight, so you can watch the shows later on demand.

And because media streaming is severely throttled during the day with this plan, daytime regular web surfing is actually treated as unlimited too.

Other Unlimited Alternatives

Subscribe to DSL/Cable at RV Parks and Mobile Home Parks: If you'll be staying places longer term, some seasonal parks and mobile home parks have cable run to the sites. Sometimes it's as easy as calling the local cable company to arrange to subscribe and turn it on.

Apple Stores: All Apple stores (and the less common Microsoft stores too) provide extremely fast free public Wi-Fi. Especially for doing OS X or iOS operating system updates, these stores are a great place to go to handle some heavy data downloading needs.

Libraries: Many libraries offer desks with free Wi-Fi or even wired ethernet connections available, and power outlets too. For a quite stress-free place to get some major downloading done, nothing beats spending an afternoon hanging out at a public library

Co-Working Spaces: Co-working spaces can be found in many urban areas, and they often make it easy for travelers passing through to rent some desk space for a day/week/month – complete with access to their high speed unlimited internet.

Thinking Outside the Box

Cellular, Wi-Fi, and satellite are not the only ways to get online while enjoying life on the road. If you are willing to get a little creative or experimental, there are a few other less obvious alternatives too.

Particularly if you are willing to get a little flexible with what "mobile" and "internet" means, you may have more options than you ever imagined.

Amateur Ham Radio – Email & Internet

If you're a licensed ham radio operator with the right equipment, you can get access to Winlink (www.winlink.org).

This service allows you to send noncommercial emails over the amateur radio frequencies – which can be useful if you're out in the middle of nowhere with no other options.

The important thing about amateur radio is to remember that it is strictly for amateur use – it is against the law to use the amateur radio bands for any commercial purpose (including checking work email) or to transmit any encrypted data, so doing online banking is out.

It also means that for the most part you need to be comfortable figuring things out on your own – commercial companies aren't going to hold your hand and coach you through the process of getting online. But if you are interested in pursuing the ham hobby, there is an incredibly vibrant and experienced online community very willing to mentor newbies.

If tinkering with radios turns you on, this can be a great free way to get online. Even with the inherent limitations, as a ham you will be able to manage some very basic data connectivity on the go, wherever you are.

And if you have a big enough ammeter radio rig and the appropriate antennas, you can usually get connected from just about anywhere on Earth.

An additional ham resource we're aware of is Broadband-Hamnet (hsmm-mesh.org/) – an experimental, high-speed, wireless broadband network that uses Wi-Fi networking gear with custom firmware to build a self-configuring mesh network. If you are in an area with other users, you can be part of creating a widespread network that stretches much further than a single user alone could ever communicate.

If you want to get involved with amateur radio, the place to start is the ARRL (www.arrl.org) – the national association for amateur radio.

Cable / DSL / WISP Installation

Being mobile doesn't necessarily mean having to use only mobile internet!

If you're planning to be in one spot for a while, sometimes hooking directly up to fixed, wired cable or DSL is a possibility.

RV parks and even mobile home parks with RV spaces that cater to long-term residents and permanent dwellers sometimes already have cable pulled to each site, and all it takes is contacting the local cable or DSL provider to get service switched on and have them bring you the necessary modem.

The park may not advertise this as an amenity, so you probably need to ask. Start by looking for parks that offer cable TV or seasonal rates.

Depending on the provider, the costs to get started are very reasonable.

Since there are often not any long-term contracts required to get service, you can cancel after just a few weeks or month or two without penalties.

And you can usually rent the equipment for a few dollars a month, instead of buying it. But if you find yourself signing up for cable internet often, many providers utilize the same modem standard – so it may be worthwhile buying a cable modem and keeping it onboard for quicker activation.

The advantage of going with cable or DSL is gaining access to fast and essentially unlimited internet. After rationing out bandwidth on the road for months on end, spending a few weeks drinking from the firehose can feel incredibly decadent and absolutely awesome!

Temporarily embracing a wired lifestyle can sometimes be very worth it for us bandwidth junkies – perfect for hyper-focusing on a work project before heading back out on the road!

Caution: When you rip yourself away from unlimited fast bandwidth again, you may whimper.

WISP Access (Wireless Internet Service Providers)

Particularly in the mountainous West, a WISP may be an option for temporary fixed-location service, even in places beyond where cable and DSL providers reach.

A WISP is a wireless internet service provider – and these companies have sprung up in many communities to fill the demand for faster-than-dial-up home internet service. The WISP providers set up transmitters on local high points and then install compatible broadband receivers on the roofs of local customers.

Since the WISP doesn't need to run new wires, if you have a view to the right mountain or hillside, you might be able to get a local WISP to offer you fast unlimited service – even if you are in a remote boondocking spot miles from the nearest cable run or phone line. It all depends on line of sight.

To find out if WISP might be an option in your area, check local advertising periodicals, signs in grocery stores or laundromats, the Yellow Pages, Google, or talk to local computer repair professionals for leads.

Because a professional installation is usually required, WISP service is not usually appropriate for short-term stays.

Co-Working Spaces

Co-working spaces are office setups for independent workers to base themselves out of. They offer amenities like a desk, connectivity, and meeting rooms. Sometimes they can be far more productive spaces than trying to get work done from a cafe, as everyone around you is also working.

Co-working spaces are now located all over the country, particularly in larger cities. Some of them offer short term rental options including hour and day passes, or even monthly space. These can be an ideal option for a mobile worker to setup a work base camp for a bit.

Start with resources including CoWorking (wiki.coworking.org), ShareDesk (www.sharedesk.net) and Wherever Worker (www.whereverworker.com) to located co-working spaces available.

Concierge Services and Personal Assistants

Sometimes the best way to get stuff done online is to let someone else do it for you.

Think about how long it might take you to research a topic or make a reservation with a web connection that bounces up and down more often than a yo-yo, and a sporadic voice signal that limits conversations to "Can you hear me now?" and "Wait, what was that?" and "#%$*@! stupid phone!"

What if instead you had someone with a rock-solid internet connection willing to do your bidding on demand, and who will get back to you later with the results or needed information?

It may seems counterintuitive at first, but consider these options...

- **Tap into Online Community** – Rather than fighting with a slow connection for hours researching options, try instead to get online just enough to post a request for advice and information to your Facebook friends, your blog, or to to a relevant forum you participate in.

  Go for a hike, return a few hours later, and you often may find yourself with a wealth of information waiting.

  This is a great way to get routing recommendations or general technical advice. And it sure beats hours of cursing at web pages slowly loading! Just be sure to carefully weed out the advice you don't need – you will find everyone has an opinion, and not all are appropriate for your situation. And, of course, return the favor for others when you have a good internet connection and someone needs a little information.

- **Friends/Family** – If you have specific requests and tasks that need to get handled and your lack of good connectivity is making you pull your hair out, don't be afraid to get a message out to a friend or family member asking for help, particularly if you are dealing with an urgent or emergency situation.

  And don't forget to thank them. Maybe by sending a photo from your travels or a gift?

- **Credit Card Concierge** – There is a chance that you might already have a personal assistant on call and not even know it! Some credit cards (such as Visa's Signature cards and many AMEX cards) offer a concierge perk, and it can actually be surprisingly useful.

We once found ourselves camped out in the boonies with minimal data and voice service, and urgently needing a dentist. Rather than struggle with limited signal to find a dentist who could see us on short notice, we got a call through to Visa and tasked them with finding a nearby dentist willing to see an urgent new patient that very morning. Visa called around and even made us three tentative appointments.

- **Online Virtual Assistant Services** – If you have tasks that are more complicated than a free concierge can handle or you want a concierge who "knows you," there are a lot of online personal assistant and concierge services that you can subscribe to.

Check out Ask Sunday (www.asksunday.com/), which assigns you a "Dedicated Assistant" with monthly plans starting at $119/mo, or an on-demand service for just $29/month. Fancy Hands (www.fancyhands.com) is another service which handles smaller requests in packages as low as $25/month.

- **Dedicated Personal Assistant** – And finally, some nomads have found it worthwhile to actually have someone on staff. If you are running a business and need to keep up appearances of always being available, having someone who answers the phone, triages replies to emails, and handles making reservations for you can buy a lot of flexibility and freedom.

Having a fast and reliable assistant can ultimately minimize your needs for staying close to fast and reliable internet.

Routers – Bringing It All Together

Routers serve as the central conductor on any network – acting as a gateway between the Local Area Network (LAN – in other words, YOUR devices within your RV) and the Wide Area Network (WAN – 'the internet').

If you want to have more than one device taking advantage of a single upstream internet connection, or you want to connect your local devices together to share files or functionality – you need a router.

Typical home routers often connect to a cable or DSL modem for the WAN uplink, and create a local Wi-Fi and wired ethernet network for all the local devices in a home to connect to and share this upstream connection.

But in an RV – cable modems and DSL lines are rarely found.

Instead – upstream connection options are always changing between cellular modems and public Wi-Fi hotspots – which takes a special kind of router to be able to interface with. Typical home and office routers are just out of their element in a mobile environment.

The most basic cellular router is a Mobile Hotspot or MiFi or Jetpack – as discussed in the Cellular Gear chapter. A smartphone creating a "Personal Hotspot" is doing the same, acting as a router and sharing its cellular connection with other nearby devices.

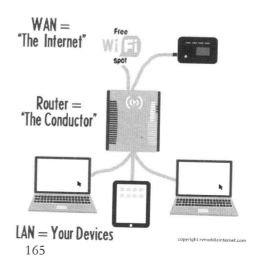

WAN =
"The Internet"

Router =
"The Conductor"

LAN = Your Devices

Routers – Bringing It All Together

If you need more capability or flexibility than a mobile hotspot can provide, more advanced router options are out there – allowing you to connect to both cellular and remote Wi-Fi networks as upstream WAN sources, and making it easy to change between upstream connections too. You can even have WAN inputs like satellite, cable, or DSL should you have access to them.

Some routers even let you connect & combine multiple WAN networks (ie. two cellular networks, or a cellular and campground Wi-Fi) at the same time for increased speed or reliability.

The core features that set RV-friendly routers apart from home routers is support for at least some of the following features:

Embedded Cellular Modem: A router with a cellular modem integrated in to it can use a SIM card and an active data plan to connect directly to a compatible cellular network.

USB Cellular Modem Support: A router with a USB port that can control a USB cellular modem or tether to a cellular mobile hotspot allows for a cellular connection to be shared. But the number of compatible USB devices may be limited.

WiFi-as-WAN: A router with this feature can connect to an external WiFi network upstream (such as a campground Wi-Fi hotspot), and at the same time create a private Wi-Fi network downstream for your local devices to share that connection. This allows all of your personal devices to always connect to the router, and lets the router worry about what is the connection that best works at your current location.

External Antennas: Getting an antenna out a window or up on your roof can drastically increase your ability to bring in a solid connection from afar.

Long Range Wi-Fi Radio: Wi-Fi is generally considered a short-range wireless technology, and most Wi-Fi devices are only designed to communicate a few hundred feet, at best. But if your goal is connecting to and sharing a remote campground or other public Wi-Fi hotspot, it helps to have a more powerful radio designed for the job.

Flexible Power Inputs: Many RVers want to be able to optimize for 12v power to run off their house battery systems when off-grid. A hallmark of mobile router is the ability to run off 12v directly.

Do You Really Need a Router?

The most basic function of a router is taking an upstream network connection and sharing it with multiple downstream devices over either wireless Wi-Fi or wired ethernet.

If you never intend to share a single connection across multiple devices, you do not need a router. A simple USB modem plugged directly into a laptop or a direct connection on your phone or tablet can actually save you a ton of headaches.

And if you only have a few devices that you want to share a connection, you can probably just use the routing capabilities built into the personal hotspot feature on your smartphone or a Jetpack.

If, on the other hand, you have an entire collections of devices that you'd like to get online, potentially via multiple upstream connections, as well as enabling your devices to talk to each other – you almost certainly would benefit from having a more-powerful router sitting at the heart of your network.

In our case, even our lightbulbs and bathroom scale are Wi-Fi enabled, and a router is an absolute necessity!

Router Selection Tips

Some nomads use traditional home wireless routers, such as the Apple AirPort or any of the vast range of routers sold by LinkSys, D-Link, Netgear, and many others.

These home routers, however, do not have any built-in support for cellular data connections, making them substantially limited for mobile use.

Some general consumer routers and Wi-Fi range extenders support WiFi-as-WAN functionality and a whole raft of other features, but they are all designed for extending Wi-Fi networks where you control both ends of the connection – a situation that is hardly ever true for an RV traveler.

You will save yourself some headaches by going with something designed with the mobile user in mind.

Keep Current Alert: The models offered are constantly changing, and we track & compare them in our member guide to mobile routers, found at: www.rvmobileinternet.com/resources/cellular-wifi-as-wan-routers/.

Cellular & WiFi-as-WAN Routers

Though it ends up being a more complex setup than a MiFi, often there are advantages to using a specialized router that can bring together multiple upstream connection options.

The key feature to look for is direct support for controlling cellular modems and for a feature known as WiFi-as-WAN – the capability to simultaneously create a local private Wi-Fi network while also connecting upstream to a public Wi-Fi network.

This scenario is common for RVers connecting to campground Wi-Fi, a friend's hotspot, or a network hosted by a merchant.

A mobile router may also be able to juggle other upstream data sources, such as a cable modem or satellite connection – and you should look for failover features that define how these connections are prioritized and automatically switched between.

One common configuration is to set up your router to automatically connect via WiFi-as-WAN to any open Wi-Fi network, and failing that, to automatically switch to a cellular connection. This way the free and unlimited connection becomes the priority.

The nicest thing about using a router like this in the heart of your network is that it keeps things simple on all your client devices. You just need to configure all your devices to point to your router's hotspot, and when you change locations you do not need to reconfigure any of your devices or reenter any new passwords.

Here area some of the current router manufacturers with products suitable for RV use:

> **WiFiRanger:** WiFiRanger (www.wifiranger.com) is a small company focused on providing Wi-Fi/cellular routers for the RV market. Because of this focus, WiFiRanger has invested a lot of effort into keeping things as simple as possible for nontechnical users, while attempting to solve pesky problems like getting the router successfully past campground login pages.
>
> WiFiRanger has various models designed for both interior use as well as outdoor long-range CPEs designed to be mounted on the roof – and when the indoor and outdoor units are paired they work together automatically via a single control panel. For most RV users with serious Wi-Fi needs, the WiFiRanger is a great all-around choice.

168

Pepwave: Pepwave (peplink.com) is primarily focused on the enterprise market and produces a whole range of advanced and extremely capable routers. They do have some affordable consumer-level products that many RVers rely on, with a reputation for rock-solid reliability.

Cradlepoint: Cradlepoint (www.cradlepoint.com) has been backing away from the general RV and consumer-friendly mobile-router business to focus on higher end markets. Cradlepoint has a whole range of high-end mobile-office gear that remain a top-notch choices for those with the most demanding needs.

MoFi: MoFi (mofinetwork.com) is a small Canadian company that creates a consumer level mobile routers with many advanced features integrated – including router options with integrated cellular modems.

Advanced Configurations

Depending on your connectivity needs and tolerance for or enjoyment of network hacking and configuration juggling, you can build up an even more complex router configuration.

One good place to start is by checking out DD-WRT (www.dd-wrt.com), an open source alternative router firmware project that lets you tweak, configure, optimize, and automate router functionalities.

If you have a supported router model and are willing to void the warranty by replacing the factory programming with something more complex, geeks can have a lot of fun here. Some routers can now be purchased with this firmware pre-installed.

But even serious geeks often prefer to have something more simplified, tested, and supported.

You can get many of the same advanced capabilities and more by going with a high-end, commercial-grade router from Cradlepoint, Pepwave, or others.

One feature common in higher end routers is load balancing – the capability to use multiple upstream connections simultaneously to spread out connection demands, balancing usage.

Connection bonding takes load balancing even further – creating a VPN connection that combines multiple upstream paths into a single faster virtual connection. For connection bonding to function, you need to have a server reversing the process at the other end – usually another router or server running software from the same manufacturer.

A well-configured router with load balancing enabled can actually take into account network speeds, data caps, and costs – automatically attempting to optimize your usage across multiple carriers.

Pepwave even has routers with integrated cellular modems featuring multiple-SIM slots, allowing the router to automatically switch to a secondary account when all the data on the primary account is used up.

Higher end routers also often support integrated VPN functionality – allowing your entire mobile network to virtually appear local to a remote network. If you are working remotely for a larger corporation, the IT staff may insist on this sort of configuration.

Wired vs. Wireless Speeds

Plugging your fancy laptop into a wired network jack may seem like a step backwards into the primitive pre-wireless age, but especially if you have multiple computers talking to each other (and not just upstream to the internet) it can actually end up making a lot of sense.

Consider – even if you have a 5GHz 802.11n wireless network, the maximum theoretical speed is usually at best 300Mbps – and in less-than-ideal conditions, the speeds will be much, much lower.

Meanwhile, a wired gigabit ethernet network can run at a full 1,000Mbps speed no matter what, in both upstream and downstream directions simultaneously.

Since we have our backups and media stores on a network attached storage (NAS) drive, having a wired network means that these drives are reachable from the other computers nearly as fast as a local hard drive would be.

If you do build a wired ethernet network, make sure to seek out hardware that supports gigabit speeds: "Fast ethernet" can barely outrun 802.11n Wi-Fi. You will also need a gigabit switch, which can be had cheaply.

Don't rely on the ethernet ports in your router, though – a lot of mobile routers seem to not (yet) have gigabit switching built in.

Apple laptops have always supported gigabit ethernet speeds, but PCs are more hit-or-miss. Check your specs to be sure the wired connection will be faster than the Wi-Fi.

Managing Mobile Data Usage

A mobile internet connection benefits from careful management.

Unlike typical home connections, which are uncapped and unthrottled and rarely changing, mobile internet connections vary in quality and speed as you travel and are always at risk of running into potentially expensive overage charges or frustrating speed throttles.

A little effort spent keeping tabs on your connection speeds and usage can go a long way towards maximizing your online experience.

Tracking Your Data Usage

To determine how much data you use, and avoid surprise overages or accidental account suspensions, you need to keep on top of your data usage. Most internet providers give their customers a way to check usage directly with them – either via a customer login account online, an app, or sending a text message on smartphones. Check with your carrier for instructions on how to do this for each of your devices.

Set it up, and get in the routine of checking each of your internet sources throughout the month to make sure you're staying within your quotas.

Some devices also have built-in usage tracking right on the device. Many modern mobile hotspots make this really easy and display your usage and limits right on the control panel screen, or even via a companion smartphone app. The device retrieves the carrier's report of how much data has been used so far in the billing period for display.

If you don't have an easy way to tap into your carriers usage metering in real time, you'll need to track it on your own.

Tracking your own independent usage can also help you figure out if you have a rogue program sucking up bandwidth – such as a sync to the cloud or a large update downloading. And if there's an unexplained spike in usage on your carrier's reporting, you can go back to your logs and see if it's accurate or perhaps an erroneous report that you need to refute with the carrier.

There are multiple ways to independently track your usage.

On-Device Usage Tracking Tools

You can install a tracking tool on every internet-using computer or device to track usage and monitor speeds.

For Windows, check out:

- TripMode (www.tripmode.ch)
- NetWorx (www.softperfect.com/products/networx/)
- BitMeter II (codebox.org.uk/pages/bitmeter2)

For Mac, check out:

- Activity Monitor (pre-installed with OSX) can provide data usage for the network since your last reboot.
- TripMode (www.tripmode.ch)
- Little Snitch (www.obdev.at/products/littlesnitch/)
- iStats Menus (bjango.com/mac/istatmenus/)

The advantage of on-device tracking is that these programs can often track instantaneous usage per-app, helping you isolate the hogs.

It is difficult however for these programs to tell the difference between local area network traffic (such as between two computers on your network) and internet traffic that is using up your cellular data.

And the big limitation is that you need to total up your overall usage across all your devices manually, and these tools have no way of measuring off-device usage – such as a video game console's data consumption.

Router-Based Usage Tracking

To really know how much data all of your devices are using, you need to funnel all the usage through a single point where it can be monitored.

That's when a router comes in super handy.

Most routers have some sort of usage tracking capability – but how useful and well implemented these features actually are vary greatly.

Pepwave products in particular have some impressive usage-tracking tools that can even help you isolate which devices on your network are the most hoggish, graphing usage by device over time.

WiFiRanger (pictured below) also includes a a bunch of truly robust usage tracking, monitoring, and load balancing features that even allow you to set data usage allowances by device – which can help prevent rogue downloads, monitor usage by kids in the household, or just better fine tune your bandwidth management.

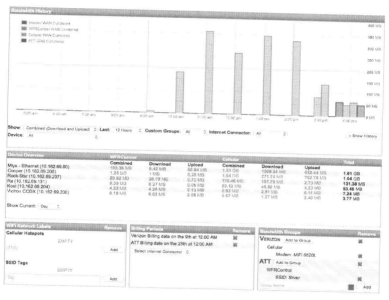

Android Usage Tracking

If you are using an Android device as your upstream connection – the Android OS has had a very powerful built in cellular data usage tracker since OS release 4.0 (Ice Cream Sandwich). This tool is found under the "Data Usage" section of the settings menu.

Not only can you see usage by app, you can set the dates of your monthly billing cycle, and even set warning and cutoff limits to prevent overage charges.

There are also dozens of other data tracking tools available for Android.

iOS Usage Tracking

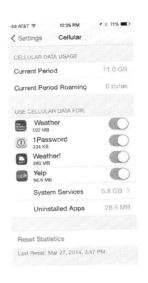

Apple has more primitive built in data usage capabilities than Android, but you can still track overall and per-app data consumption under "Cellular Data" on the Settings menu. You can even block badly behaving apps from being able to use cellular data with just a tap.

In case you are looking for it, the data used by the Personal Hotspot feature is hidden beneath the System Services selection.

The catch with the built in iOS data usage tracking is that you must manually reset the statistics periodically, making it hard to keep track of usage coinciding with your current monthly bill.

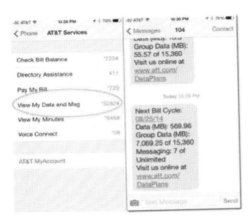

To see your current monthly usage, hidden at the bottom of the Phone menu on the Settings app is a place for your carrier to add special features. AT&T has used this to provide a button that will request a text message with an update you on usage in your current billing cycle, showing both your personal usage as well as the total used and remaining data on a shared plan.

Tips for Minimizing Data Usage

When you're on mobile data caps, you have to learn some tricks to minimize your data usage. In particular, LTE data can easily get burned quicker than most folks realize is possible if you are not careful.

Here are some ideas to minimize that amount of data you burn through:

- Unless you need the speed, you may be able to force your device to connect via 3G by disabling LTE support in the device settings. For a lot of basic tasks, 3G is fine, and the slower speeds will help minimize your data usage. On the other hand, 3G networks are starting to get pushed to the back burner and performance may suffer more than you can bear. Also - a lot of devices no longer have the option to force your connection to 3G when LTE is present.

Managing Mobile Data Usage

- Be very careful when you load a page with video on it. If it autoplays, it's very likely caching faster than you can watch it. This means if you click away partway through the video, you've already spent the data – regardless if you actually saw the whole thing. Even if you click "Pause" or "Stop," it still caches in the background.

- Another video gotcha: Sites will often automatically play back HD video if you are on a fast connection, rapidly burning through data!

- Make sure your bookmarks to sites you visit frequently go directly to the actual page you start your experience with. You may have a bookmark to your bank's website, not to where you log in. Or a link to a forum you participate in – but the link goes to the front page, not the listing of recent postings you like to start with. No sense wasting bandwidth just to load a welcome page!

- Make sure you have auto-downloads of system and software updates turned OFF if possible. Save those for when you have unlimited internet! A big OS update might be gigabytes in size, in one download using up all your data for the month. Wait and download these later.

- Windows 8 & Windows 10 support labeling Wi-Fi networks as "metered connections" where Windows will not attempt to download updates. Make sure to do this for all data-limited connections - especially on Windows 10 where you can not otherwise opt-out of software updates!

- Watch out for iOS updates! In the past iOS devices have been known to automatically download OS updates when on a Wi-Fi connection, with no way to manually opt-out other than staying disconnected from Wi-Fi. If your Wi-Fi is provided by a cellular hotspot, this unwelcome update could push you into overage territory when a major new release comes out.

- Pause auto-syncing to your cloud-based backup services, like DropBox, when on limited connections. Some, like CrashPlan, have features to disable backups when connected to certain Wi-Fi networks.

- If you are subscribed to podcasts, TV series, or any other periodical content through programs like iTunes, be extra vigilant that you are not auto-downloading new episodes in the background.

- Be aware of programs that sync or upload data - especially ones you're not actively using. Check to make sure iPhoto isn't uploading to iCloud or Flickr, iTunes isn't auto-downloading app updates and podcasts, Adobe Creative Cloud product updates are off, and more. Even if you think you have the settings tuned just right, things can easily get messed up – sometimes causing repeated attempts at partial syncing and uploading.

- If you have multiple computers and devices, make sure that they aren't duplicating downloads. If you are subscribed to a podcast that you want to download new episodes of, make sure that you are doing so on only one of your devices. No sense having three copies of something!

- Run an ad blocker in your browser to avoid loading unneeded graphics and promotional video ads. Disabling Flash support can make a huge difference too.

- A lot of apps, unless they are specifically designed for mobile usage, are built with the assumption that data is unlimited. Even when they appear to be idle, some apps are actually burning through data in the background – downloading updates, pre-loading content that you may never even view, or even updating advertising for quicker display later. Be careful of what you leave running in the background!

- And if you're not using it, turn your internet devices OFF or disable internet on your computer to prevent background tasks from silently eating away at your data allotment.

- Game consoles, Blu-ray players, and even some smart TVs can use up a ton of data silently in the middle of the night downloading updates. Do not leave these devices plugged in to mobile internet unless you are keeping a close eye on them.

- If you are an advanced user, you can set up your own online proxy server that you route all your connection requests through. The proxy then can compress the data before it gets transmitted, potentially even recompressing graphics to be lower resolution or throwing away data that the proxy feels is unimportant.

  Some cellular and satellite internet connections actually do this for you automatically in the background – you may sometimes notice that pictures look slightly worse when on cellular because of this.

- Note, though, that a proxy server (whether your own or via your cell provider) can only compress and modify unencrypted pages. If you are visiting an encrypted site, no one in the middle can modify what is transmitted (for good or for ill).

- Android users can use Chrome's Bandwidth Management. To start saving data using the Chrome for Mobile browser, visit Settings > Bandwidth Management > Reduce Data Usage. Then simply turn the toggle to "On." From this menu, you'll also be able to track how much bandwidth you save each month as you browse in Chrome.

No matter how careful you are, you will still inevitably stumble into accidentally using way more data than you realized or thought possible.

It is tempting to blame your internet provider and to disclaim all personal responsibility. We hear it all the time – "there's no way I could have used that much data!"

Heck, we've said it ourselves when notified of high usage.

But in reality, screwed up accounting behind the scenes is rarely to blame. Very often some forensic digging finds the real guilty party. We have seen OS updates and iTunes ignore "auto download" settings, upgrades to Google's Photos upload tool silently opt-in to "backup all photos to the cloud," iPhone/iPad OS updates auto download when connected to Wi-Fi (even over cellular Jetpacks), and even an Xbox One turn itself on in the middle of the night to download a huge 4GB update with no disclosure or warning given.

If you ever accidentally blow through your data caps on your primary account, it is extremely useful to have a secondary way online to help you limp through to the end of the month when your caps reset.

Optimizing Social Media Browsing

Facebook, Instagram and Twitter now autoplay video in your timelines and default to showing high definition video.

Thankfully each service provides a way to reduce automatic data sucking, allowing you to adjust what you want to use your precious data for. However, you have to manually override default settings. And some of the settings do not apply when using Wi-Fi, as the services assume those sources are always unlimited. This is not necessarily true for RVers.

Twitter: Disabling Video Auto-Play

By default Twitter enables video auto-play – but it also provides ways to turn off auto-play both in the browser view and in the app.

On the browser/desktop view:

- Click on your User Icon in the upper right hand corner to display the Profile & Settings menu, and select 'Settings'

- Scroll down and unclick the Video Autoplay option under 'Video Tweets'

For the mobile app version:

- Tap the 'Me' tab in the lower-right corner

- Select the gear icon, then select 'Settings'

- In the General section, choose either 'Use Wi-Fi Only' or 'Never Play Videos Automatically' for Video autoplay.

Remember to save your changes!

Facebook: Disabling Video Auto-Play / HD Video

Facebook is already a time suck for many, but it's also a heavy mobile data consumer when it comes to displaying photos and videos. By default, Facebook will autoplay any videos shown on your timeline – whether you want to view them or not.

Here's how to control when you play videos.

On the browser/desktop view:

- On the very top right hand corner of Facebook, click on the little down arrow to bring up the Settings Menu.

- Click 'Settings' from the pull down.

- Then click 'Video' from the left hand menu that appears.

- This will bring up two options:

  - The first one allows you to select if videos display in SD (standard definition – which is lower resolutions, and thus uses less bandwidth) or HD (high definition, which will use a lot more bandwidth). This will be your default setting but you can manually override this when viewing videos. We recommend setting this to 'SD Only', and then if there's a video you want to see in HD select that at time of playback.

  - The second allows you to select your autoplay back settings. For mobile data users, we recommend selecting 'Off'. When a video is posted to your timeline, it will now just show as a thumbnail, and you can click on it to play.

For the mobile app version:

- Tap on the 'More' tab in the lower right hand corner.

- Scroll all the way down to almost the bottom of the menu and select 'Account Settings'

- Select 'Videos and Photos'

- Under Video Settings, turn off uploading in HD — unless you want to post high definition videos from your mobile device.

- Click 'Auto-play' to access the auto-play menu.

Instagram: Minimizing Data Usage

Instagram displays photos and does allow posting 15 second video clips. These video clips auto-play, and Instagram has removed the option to prevent this. Instead, they now have a 'Cellular Data Use' setting you can enable.

Here's how to do that from the mobile app version:

- From your account profile tab (click the 'little dude' in the lower right hand corner), select the 'Gear' at the top of the page.

- Scroll down until you find the 'Cellular Data Use' menu item, and tap it.

- From this screen, enable 'Use Less Data'.

All it does is stop the app from pre-loading videos before you scroll to them. It will cause videos to take longer to load, but they still auto-play as soon as you encounter them.

Network Speed & Quality Testing

If you are going to invest any effort in optimizing your mobile data connection, you need to have ways to measure the impact your changes are having.

Focusing on more bars is not enough. Using boosters and repeaters to improve the signal strength only tells you half the story.

To really evaluate a mobile network connection, you need to keep a close eye on your actual speed and latency testing results. Here is how you do it.

Speed Testing Services

There are numerous speed testing services and apps — these are the ones we regularly use:

- Ookla Speedtest (www.speedtest.net)

- Ookla Speedtest App (www.speedtest.net/mobile/) — For iPhone, iPad, Android, and Windows phones.

- DSLReports Speed Test (www.dslreports.com/speedtest)

- Speed Of Me (www.SpeedOf.Me)

If you ever get results that seem odd, try another service.

Understanding Speed-Test Results

You will get three results from most speed tests:

- **Latency (aka Ping):** This is the time in milliseconds it takes for a request from your computer to reach the speed-test server and to return, like the ping of a ship's sonar. The higher the number, the slower the speed.

Latencies under 100ms are good, under 50ms are great, and over 500ms begin to feel painful. This measurement is particularly important for online gaming – but any interactive task can begin to suffer with higher latencies.

- **Download Speed:** Reported in either kilobits per second (Kbps) or megabits (equivalent to 1000 kilobits) per second (Mbps). This is a measurement of the maximum speed that data is able to flow to you from the speed-testing server. Speeds over 5Mbps give a good surfing experience, and over 20Mbps will feel awesome. Speeds under 1Mbps start to make the modern internet feel slow, and speeds under 500Kbps are painful.

Download speeds have a particularly huge impact on streaming audio and video – if the speeds aren't able to keep up, you will experience stuttering, pauses, and long buffering delays. If your speeds are below 500Kbps, don't even waste any time trying to use video sites like YouTube.

- **Upload Speed:** The opposite of download speed – how fast is data able to get from you to the speed-test server. Upload speeds are almost always substantially lower than download speeds, and for many typical internet tasks upload speeds don't have a huge impact. But upload speeds are critical for video chatting and, of course, uploading large files like photos, videos, or cloud-synced backups. Speeds over 500Kbps are the bare minimum for video chat; speeds over 1.5Mbps can deliver smoother results.

Speed Test Tips & Tricks

Speed tests can vary a lot from moment to moment – and a lot of that variability may not have anything to do with your personal network connection. To get a sense for the actual health of your connection, running several speed tests over the course of 10 or 15 minutes can help you get a better sense of what average speeds you are actually achieving.

Most speed-testing sites and apps have a way to change the server – letting you select a different server to communicate with and test against. Trying different servers can help you rule out whether strange results are isolated or not.

If you are comparing usage between two devices, make sure that your speed tests are using the same server! An overloaded server can make one connection test slower than another, when in fact it might actually be faster.

And finally – keep an eye on your data usage. Speed tests work by sending large chunks of data back and forth to the server, timing how long it takes. Excessive speed testing can burn through your monthly data bucket rapidly if you are not careful.

Scoring "Free" Data

A potential big shift in how mobile networks work and are paid for is called "zero rating." Zero-rated content does not count against your data allotment for the month. The rate charged to transmit the data is thus "zero" for the end users.

Think of it like an 800 number used to be on a voice phone network.

T-Mobile's Music Freedom and Binge On free video streaming features are pioneering the zero-rated concept – with neither the customer nor the content provider paying extra for the data consumed.

A different take on zero rating involves sponsored data – where a sponsor is footing the bill for the content instead of the viewer.

Verizon and AT&T are both pursuing sponsored data initiatives.

Verizon's program is known as "FreeBee Data", and Verizon is hoping that as the program goes mainstream in 2016 customers will learn to "See the bee, click and it's free!" – with a small bee icon indicating content that is free to watch.

Verizon's go90 video streaming app is (not surprisingly) the first major service to embrace sponsored data – with go90 now free to watch for Verizon postpaid customers, with the data being paid for by the advertising.

AT&T has been slowly experimenting with sponsored data too.

A company called Syntonic Wireless (www.syntonicwireless.com) has rolled out an iOS and Android app that acts as a portal to sponsored content on AT&T – allowing for a range of sites to be visited (only through the app) without it counting against your monthly cap.

Sites that have signed up with Syntonic include Airbnb, Amazon, eBay, ESPN, Etsy, Expedia, Facebook, MLB.com, Open Table, Rolling Stone, the *New York Times*, and Yelp – with more on the way.

It is unclear how much information the sponsoring sites get handed about you in return for paying for your access.

Network Neutrality & The Downside of "Free"

These zero rating schemes have generated a lot of controversy.

On one hand, "free" is great for users fighting to stay under their monthly caps – but on the other, a lot of people view this as a genuine threat to the idea of an open internet – with nonparticipating sites left at a competitive disadvantage, and the carriers left with little incentive to bring out lower prices for non-sponsored data.

After all – the more expensive data is, the more likely customers will be compelled to seek out toll free alternatives.

It has been unclear whether or not these emerging practices may bump against the FCC's Network Neutrality rules that prohibit networks from "playing favorites" by treating some content more favorably than others.

The FCC has indicated that it will be keeping a close eye on how these free data plans continue to evolve.

Strategies for Hard Drive Back-Ups

Hard drives will inevitably fail – it is not a question of if, but when.

Phones get dropped. Tablets get stolen.

And sometimes computers in RVs can very literally 'crash' – headlong into another vehicle.

In all these cases – the loss of the hardware is often not nearly as catastrophic as the potential loss of the data stored.

Backing up critical data is one of the most important chores for any responsible technology owner – whether you are worried about critical client documents on your laptop, or irreplaceable photos on your phone.

To be safe – all important data should be stored in at least three separate places, with at least one of the copies located physically offsite in case of a total catastrophe.

This sort of best-practices behavior is hard enough to achieve when living and working in a conventional fixed-location – but for those of us on the road living with limited data connections this can feel downright insurmountable.

Here are some tips to better manage backups on the road – without blowing through obscene amounts of data doing it:

- **Cloud Storage & Backups** – Conventional online backup services that copy the entire content of your hard drives to the cloud are great if you are on a fast truly unlimited connection, but are a recipe for disaster for mobile users.

  But that doesn't mean that you can't use the cloud intelligently to your advantage.

  Rather than backing up everything, set up a shared folder with a service like Dropbox or OneDrive. You can then use this shared folder for your most critical files and for projects that you are actively working on.

  In the event of a computer crash or loss, you'd be able to do a system restore from your last full back-up you can get your hands on. In the meantime, you can get immediate access to your active files to get right back to work.

  Be careful if you are working with extremely large files though. A large video or graphical file that is auto-saved to a cloud drive every few minutes as you edit can burn through massive amounts of data in no time. But you can pause syncing while you are working on large files or copy them to your local hard drive. Small files will use relatively little data at all.

  If you do use a full-HD cloud backup – make sure that your initial seed backup has plenty of time to complete before you hit the road. Some services will even send you a blank hard drive to back-up everything you wish, and then you mail it to them to store in the cloud for you – saving tons of time & bandwidth.

- **Local NAS / External HD** – A permanently attached external HD or a network attached storage device (NAS) attached to your router can serve as a great local backup for all your data. This will let you quickly recover from accidental deletions or major malfunctions on your computers.

  TIP: To protect your drives from vibration, do not leave your NAS powered on while underway. Only more expensive SSD drives with no moving parts are immune to vibration or knocks.

- **Offline Fire Safe HD** – It is a good insurance policy to also have a weekly or monthly backup around that is kept disconnected and safely stored when not in use. A computer virus, a fire, or a thief may take out your computer and everything attached to it – but a backup hard drive stored in a fire safe may still be fine.

- **Offsite Mirrors** – To guard against the worst – it is important to leave a backup securely somewhere else, updated as frequently as you feel necessary.

  We typically leave a backup drive stored with friends or family in various places around the country. Other nomads mail their backup drive to their mail forwarding service – leaving it stored there until they are ready to request it again for updating.

- **Don't Forget Your Gadgets** – Even people who religiously back up their computers often forget about the other tech in their lives. Make sure that your phone (and all the photos your take!) are being backed up regularly – whether to the cloud via services like iCloud Backups (but watch out for data usage), or to your computer by remembering to do an occasional manual sync.

Entertainment on the Go

While RVers do enjoy getting out exploring our new locales, hiking in nature, visiting with friends — when RVing is a lifestyle it's about balance.

There's no shame in enjoying watching television or a good flick, or passing some time playing games.

This is life on the road, not a vacation. We are RVing, not just camping and spending all our evenings roasting marshmallows over a campfire.

There will be bad weather days, or days you're not feeling well, or just days you're overwhelmed exploring yet another location (yes, it happens). Unwinding after a day of work, exploring, and socializing in front of the tube isn't a crime.

Many online entertainment options have evolved to consume a lot of internet data — streaming video, online gaming, or even just general web surfing. And it seems that content creators and providers are always increasing the quality of their offerings, which means they eat up even more data all the time!

If you have fast and unlimited data, like cable internet, then this is no big deal.

But for those of us managing mobile connections with capped data and variable speeds, we quickly start running into problems if we're not willing to adjust our expectations and viewing habits.

With some forms of mobile connectivity being as fast, or faster, than home-based cable services, you can use your capped data up quicker than you might think by watching just a couple movies online.

Watching it on public Wi-Fi hotspots — such as at campgrounds? Maybe.

Keep in mind, most Wi-Fi hotspots are configured to allow guests access to email and basic web surfing. Just one or two folks streaming videos over a shared connection can bring the network down for everyone in the RV park, and several folks online doing "normal" web surfing will not leave enough capacity available for anyone to stream video without stuttering.

Many RV parks have gone to limiting how much data their guests can use daily so that everyone has a fair shot at using the resource, and others specifically do not allow streaming video at all.

Even if you find a park that doesn't specifically forbid streaming or limit internet usage, please be a good neighbor and don't hog all the capacity unless you know the network is up the challenge.

If anything, limit nonessential high-bandwidth usage activities to off hours – such as overnight or while everyone is at work or out sightseeing.

Streaming Video – Netflix, YouTube, Hulu, Etc.

So often we hear from folks that their mobile internet needs won't be so demanding because they're not trying to work online or attend remote virtual classes, they *just* want to stream some movies and TV shows.

Unfortunately, streaming video is one of the most bandwidth-intensive things you can do on the internet.

An hour and half high-definition movie on Netflix can easily eat up **4.5GBs** of data! For perspective, if you have a 10GB plan on Verizon at the current rate of $80/month, you've just used nearly $40 of data. If you're paying by the GB, such as overage data on an AT&T or Verizon at $15/GB, that movie would cost you $67.50.

Suffice it to say, streaming lots of hi-res video content is just not going to be financially feasible for most – unless you happen to have a truly unlimited data plan.

If streaming is absolutely going to enhance your life, the expense and risk of buying a grandfathered-in Verizon plan or AT&T iPad plan may be a worthwhile investment for you. Or you might be interested in an unlimited on-device data plan from a secondary carrier, like T-Mobile or Sprint. Using an HDMI cable out, you can mirror the streaming broadcast to a TV screen or projector.

T-Mobile's Binge On: T-Mobile's Binge On feature revolutionized mobile video streaming by enabling unlimited video streaming of supported video services that does not count against monthly high-speed data limits.

With Binge On enabled, video is limited to 480p "DVD Quality" resolution, but you can watch as much of it as you want over cellular.

Binge On currently supports 40+ services, including Netflix, HBO, Hulu, ESPN, and Amazon Video – with more being added all the time.

The Binge On feature works with all T-Mobile postpaid plans with at least 6GB of high-speed data a month – including smartphones, tablets and hotspots. At time of publishing however, it does not work when streaming through devices like Roku, Chromecast and Apple TV. But an HDMI cable will allow you to mirror from a tablet or a phone to other big screens.

Verizon's go90: Though it is a vastly more limited offering than Binge On, Verizon has made its go90 streaming video application free to watch for all Verizon postpaid subscribers. The catch is that you must watch go90 content on your phone or tablet – Verizon is currently actively blocking every option for outputting video out to an external larger screen.

If you're not using Binge On, here are some tips that might help a bit with incorporating some video streaming into your travels without breaking the bank.

Netflix Tips

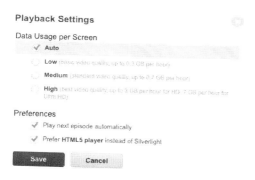

By default, Netflix will auto-adjust the quality of the video it is delivering to match your current connection speed. If you're on a solid 4G signal, this could mean very nice and smooth HD video, which will rapidly deplete a data plan. You can force Netflix to show you content at lower quality levels so that you're using your data slower.

To select a setting that works best for your internet plan, navigate to the Your Account (movies.netflix.com/YourAccount) page and click Playback settings in the Your Profile section. Restricting data usage will affect video quality while watching Netflix – but on a smaller screen, it may not matter too much.

There are four data-usage settings to choose from; here's their estimated usage:

- Low (0.3GB per hour)

- Medium (SD: 0.7GB per hour)

- High (HD: 3GB per hour, Ultra HD 4K: 7GB per hour)

- Auto (adjusts automatically to deliver the highest possible quality, based on your current internet connection speed)

Switching to low-quality from high-definition video will allow you to watch 10 times the amount of video for the same bandwidth. But, of course, the video quality may be so low that it's not worth watching on larger screens or for big-budget movies with special effects. We find the Medium setting hits a sweet spot of data usage and quality for most of our viewing.

YouTube Tips

YouTube will default to playing the best quality video for your connection speed but does provide a single setting to opt out of playing HD video.

Go to your YouTube settings and select "Playback," and then select

Playback

Video playback quality

○ Always choose the best quality for my connection and player size
Always play HD on fullscreen (when available)

● I have a slow connection. Never play higher-quality video

that you have a slow connection. This can help you automatically save some bandwidth, especially since YouTube auto-plays videos as soon as you hit a video's page.

YouTube has also added the ability to adjust playback settings while viewing videos. Just click on the Settings tool in the lower right hand corner of any video to bring up the options, and then select the playback setting you'd like by clicking 'Quality'. This is handy if you'd like to opt into splurging your data on a higher resolution playback for just a single video.

For further control, you can install plug-ins into Firefox and Chrome browsers that force a specific definition of video for playback. For Firefox, check out "YouTube High Definition" and for Chrome check out "Auto HD." Both will allow you to downscale video playback as low as 240 pixels.

Amazon Video Tips

Amazon Video streaming will automatically adjust based on the device you are using to view content and your current internet connection.

When viewing on a mobile device via their app, you can go to Settings and select 'Manage Data Usage' (for iOS) or 'Watch Videos on a Mobile Network' (for Android). From there you can select your preferred video quality to manage your bandwidth. You can use 'Wi-Fi Only' to disable video playback unless your device is connected to Wi-Fi.

For streaming to a browser or other device, the only control you have is in what format you select, such as SD, HD or UHD.

For some Amazon Prime content, Amazon now allows you to download the content while you have access to an unlimited internet source to watch offline within a set amount of time. This is a handy way to stock up before heading away from free Wi-Fi or ample internet connectivity.

Hulu Video Tips

Hulu defaults to adjusting video playback based on your current internet connection. However, if you click the Settings Gear while using their video player, you can manually select between low, medium, high and HD playback

Streaming Devices Tips

- **AppleTV / iTunes:** Go to Settings → Apps → iTunes Movies and TV Shows → Video Resolution. And then choose HD or SD (lower resolution to save bandwidth).

- **Roku:** Using the Roku remote, go to the Home screen. Use the following sequence to access the playback menu: Home 5 times, Rewind 3 times, Fast Forward 2 times. From there, you can select the speed you want playback to be at (the lower the less bandwidth used).

- **Chromecast:** Download the Google Cast extension for the Chrome browser. Click on the Cast icon in your browser bar and select Options. A page will open where you can adjust select from Standard (480p), High (720p), or Extreme (720p).

- **Amazon Fire TV:** Go to Settings > Display & Sounds > Display and open the Video Resolution section.

For other video services that you might like to utilize, check their FAQs and user-settings pages to see if they offer a way to scale back the quality of the video they deliver to you.

Alternatives to Streaming Video

Instead of streaming content, here are some ideas for getting your media fix elsewhere.

- **Rent DVDs & Blu-rays** – Netflix isn't just for streaming! The discs-by-mail service actually works amazingly well for RVers who stop in places for at least a few days at a time, as you can update your shipping address frequently. If you're going to be staying for a few days in a location that has mail delivery, that's generally enough time to ship your current disc back and get a new one shipped to you. Netflix stocks a very wide selection of discs – including TV series and documentaries.

  RedBox kiosks are also very handy for renting new releases on disc for the evening, and they're located all over the place – Walmart, 7-11s, drugstores, and more. You can rent a disk at one kiosk and return it to another down the road – all while only paying a low nightly fee, no membership required.

- **Rip content to hard drive** – While it's illegal to rip content to distribute, making a backup of it for your own personal use is in a legal gray area. We keep our legally obtained DVD collection in storage, but have ripped off-site backup copies of stuff we wanted to take with us to a hard drive. This gives us ample content to watch when we have no other way, without taking up valuable physical storage in our RV.

- **Download when you have Wi-Fi** – When you have access to precious unlimited bandwidth, that is the time to stock up on content. We'll often buy a season of a favorite network TV show via iTunes and download it to have around. Amazon Prime has now also enabled offline downloading for some of its content. Always have extra hard-drive space handy for this!

- **Buy TV series on disc** – For series that we follow but don't care if we're watching the current season as it is aired, we'll buy the seasons on DVD/Blu-ray. Generally we buy used off of Amazon. When we're done, we'll sell them back online. This has become more difficult on Amazon.com, as they have put in additional restrictions for resellers – but we've found that selling a set of seasons on eBay can net decent results. You can also take them to pawn shops or used-media resellers, and turn them in for a little cash. Or, one our favorite methods, exchange with fellow RVers for a series they have. Note: You'll have to avoid spoilers for the current season from your family and friends on Facebook and Twitter!

- **Tuner & DVR setup** – There are TV tuners that can attach to an external TV antenna and you can pick up local stations on your computer. They comes with software to turn your hard drive into a digital video recorder (DVR), so you can record content at specified times. This is an easy way to multi-task your computer setup, without investing in TVs, antennas, and separate DVR equipment. Some examples · include: WinTV-HVR, HDHomeRun and EyeTV Hybrid.

- **Public TV Viewing** – Instead of trying to stream live events or major premieres of TV shows or sporting event, check around for local pubs that might be broadcasting. Sometimes it's fun to watch a major event with a group of fellow fans, and share a brew while you're at it! Or how about an old fashioned movie night date at the theatre to get the full effect? Did you know some movie theaters will even give you permission to park overnight in their lot in your RV?

Also keep in mind that the cost of paying by the GB to watch a movie might be less expensive and/or easier than going out to a movie or bringing in a rental disc (once you factor in time and fuel). Sometimes it's just worth using up spare bandwidth at the end of the month to treat yourself to some streamed content.

TV Antennas & Satellite TV

Many RVs come with a TV antenna built in that can pick up local stations wherever you roam. This is to referred to as OTA or Over the Air. You may find that this is good enough for keeping on top of local news, your favorite broadcast TV shows, and weather alerts. However, you will find a lot of variability in broadcast quality and variety, depending on how close you are to major towns and how strong of an antenna you have.

If televised content is important to you or you want more stations than is available OTA, you can subscribe to satellite dish services. Satellite has the advantage that if you can point your dish setup to the southern sky, you can watch television from wherever you are, you can get premium stations, and your channel numbering stays pretty consistent as you change locations.

Both Dish Network and DirectTV have options that you can take with you on the road – and which will work better for you depends upon your desired offerings for packages of channels, pricing, and contract terms.

Of particular note is how each handles local stations as you travel across the country. The major broadcast channels you receive will be dependent upon the service address on your account and if you're still within range of the beam serving that location. As you change locations, the providers will let you update your service address so you can get channels from your new location.

Dish also offers a DNS (Distant Network Service) that gives you major network programming that is not based in any one particular locale, if you prefer to not keep updating your address and perhaps depend on over-the-air TV antennas for local news.

You'll also have to consider what kind of hardware setup you want: portable or roof mounted.

- A portable system will allow you to park your RV in the shade under trees, and still get your satellite dish a clear view of the sky, or not worry as much if your campsite is perfectly aligned for satellite access. But this will require setup and takedown at each stop. If you don't plan to move around too often, this may be a great compromise.

- A roof-mounted system will require your RV be to be parked where your dish can see its satellite, which may not always be possible – but the setup significantly reduces your setup time, especially with an automated system that doesn't require manually retracting a dish. If you'll be moving around a lot, this has some definite advantages, as even overnighting will still give you satellite TV.

You'll also need to consider if you want a system that you have to manually aim each time you set up, which tends to be cheaper, or an automated system that can find the satellite on its own, which of course will be much more expensive.

Gaming on the Road

Online gaming demands a lot out of a network connection – particularly action games and first-person-shooters.

But to the surprise of many, online console and PC gaming over mobile data connections is actually very doable, if you are careful to manage your data consumption and overall expectations.

Online Gaming: Performance

Online games thrive on having a reliable low-latency connection, which is a measurement of the round-trip time between your computer and the gaming server. A home cable or DSL line usually has latencies well under 50ms, and this is what many games are designed for. A 4G/LTE connection can begin to get into this ballpark, with latencies between 50ms and 120ms. 3G connections are slower, but for some games the latency will still be plenty playable. Satellite, on the other hand, does not have game-friendly latencies. With lags of 1000ms or more, your character will be dead before you know what hit you.

A lot depends on the type of game. Massively multi-player online role-playing games (MMORPGs) and driving games tend to do really well with higher latency connections. Real-time strategy games are more challenging as latencies increase, as are first-person-shooters. The type of game that will really suffer even on an LTE connection are twitch fighting games that depend on precise timing for move combos. If you want to unleash that sort of pounding, you'll need to look for an alternative connection.

The other thing you will need to be prepared for are glitches and dropouts. A cellular connection is never going to be as reliable as a hard line, and inevitably some radio interference will someday happen that turns glorious victory into agonizing defeat.

Online Gaming: Configuration Issues

Some games will work without any special configurations – but just like with some home connections, some games may require special router configurations to be able to host multiplayer games. If you hit a dead-end, 3GStore.com actually offers a free gaming tip sheet and tech support to customers who have bought a mobile router from them, and they also sell this tip sheet directly to non-customers.

Online Gaming: Data Usage

The biggest fear many gamers have is that online gaming will burn through their entire monthly data bucket in the blink of an eye. And this is a very legitimate fear!

But in general, most games are not data hogs. 3GStore.com tested a range of games, and found that most games burned between 50MB and 75MB an hour, even when using in-game voice chat features.

But you have to be very careful to keep on top of your data usage!

Games and gaming consoles are not designed to respect data caps, and it can be hard to keep usage under control. Many games will automatically install game updates and downloadable content, often without even giving any notice.

A downloadable content patch full of new levels may burn through gigabytes! One of our readers was able to use bandwidth-monitoring tools to catch his Xbox One in the act – waking itself up at 2 a.m. and downloading 4GB of data without asking for permission to do so, and with no notice given after the fact!

To be on the safe side, keep your gaming consoles disconnected (and unplugged!) when you are not actively using them, and keep a close eye on your usage. And you may find that there are some games that are just too piggish to play on a capped mobile connection.

Online Gaming: Teams & Guilds

Being part of a raiding guild, clan, or team is going to be a challenge for mobile gamers. You do not want to be in the position of letting your friends down when you have such variable connectivity and always changing circumstances. As a mobile gamer, it is better to focus on games where you can play on your own time schedule, not tied to a team.

Online Gaming: Alternatives

Hardcore PC and console gamers often look down on mobile gamers – but there are actually some pretty amazing near-console-quality games out now for mobile devices. And the best part – unlike console games, most table games are optimized for the realities of mobile gaming. You may find that you can fully get your online gaming fix without ever firing up a console.

If you absolutely need your PC, PlayStation, or Xbox fix, seek out local gaming and/or comic book stores in your travels – many have gaming rooms with fast Wi-Fi setup for gamers to come in and play.

This can be a great way to meet and play with local gamers and get some intense gaming in without needing to worry about latency or data caps.

Get Out!

You'll probably also encounter a good number of folks who tell you not to fret so much over your TV, movie, or gaming preferences.

After all, you're RVing and should just get outside. Why not explore and go on a hike?

If you're living on the road any substantial amount of time, it's about balance. If you enjoy a good movie or keeping up with your favorite shows or sports, there's no shame in that.

This is life on the road, not a vacation.

There will be bad weather days, or days you're not feeling well, or just days you're overwhelmed exploring another location. Unwinding after a day of work, exploring and socializing in front of the tube isn't a crime. Heck, we enjoy a well-put-together film – it's art!

But they do have a point – don't forget to get out there and explore too! If you're not able to get enough bandwidth today to stream your favorite show, maybe it is time to go out for a hike, read a book, or watch moss grow instead.

Crossing International Borders

RVers don't just stick to the US for their adventures.

Some cross into Canada and Mexico, and want to keep connected. Some ship their RVs to other countries to explore. Some park their RVs and travel by other modalities, exploring overseas.

General International Tips

The hurdle with international internet is not that other countries don't have plentiful options – it's getting connected to them as nonresidents who are just passing through the country on a short-term basis.

The tips offered in this section apply to both Canada and Mexico – as well as any other international travel you might embark on.

Wi-Fi

Public Wi-Fi is plentiful abroad, just like most other basic necessities of life.

You will often be able to connect at campgrounds, coffee shops, cafes, libraries, hotels, airports, and more. You will be surprised just how plentiful basic Wi-Fi can be, even in otherwise primitive countries.

When traveling outside the USA, Wi-Fi is likely going to be your cheapest and easiest connectivity solution, especially if you're only going to be in an area for a brief time when it may not be worthwhile tracking down other options.

If you're planning to mix in international travels with your RVing lifestyle, it is important to assemble your technology arsenal to include gear that is easily be portable and can be taken Wi-Fi hunting to find workable hotspots.

As with all public Wi-Fi, the usual caveats apply. Expect intermittent speeds, needing to connect in crowded public places lacking in privacy and quiet, and you will need to take precautions to keep your connection secure.

Using a virtual private network (VPN) may be a smart way to protect yourself when regularly surfing on public international Wi-Fi.

Voice

There are several ways to keep your voice phone service working while traveling internationally.

Because nearly every modern cell phone supports international roaming frequency bands, the simplest option to keep connected is to just activate international roaming with your home carrier – taking your current phone on the road. This way, your home number continues to work and will ring you wherever you go. Your default roaming rates may be expensive, but you don't have to answer every call. With your phone active, at least you'll know someone is trying to reach you, and you will have the option to answer or call back as necessary.

However, if you want to make or receive calls on a regular basis while you're out of the country, you'll probably want to avoid the default international rates your carrier is offering to keep costs better under control.

Most carriers have special international packages that offer discounted calling rates, and allow you to keep your home number active in most countries around the world for as little as $15/month.

If keeping your domestic number isn't important, local SIM cards abound in many countries.

These SIM cards will make it harder to manage incoming calls since your phone will have an international number, but many SIM cards available while traveling offer substantial discounts for calls back to the USA.

And finally, you can rely on services like Google Voice, Google Hangouts, FaceTime and Skype to handle your calls – avoiding traditional cellular calls entirely.

Cellular Data

The most important international data tip is to turn OFF roaming on your devices when you are close to international borders!

The international data roaming fees from most carrier can be extremely high – and there are many horror stories about accidental astronomical bills from inadvertent roaming.

If you do need cellular data overseas – you can avoid the default roaming rates by activating a special international roaming plan with your US-based carrier – saving a substantial amount in the process.

Roaming with your home carrier may be ideal if you're only planning a short trip, will primarily be relying on Wi-Fi, or won't be needing much data to get by with.

But for longer trips or heavier needs – it often makes sense to look into other options.

Here are some details on the four major carriers international roaming packages:

Verizon
(http://www.verizonwireless.com/landingpages/international-travel/)

Verizon offer two different options for international travel connectivity discounts:

- **TravelPass:** For a low daily rate, you can access your existing domestic plan while traveling internationally – including talk, text and high-speed data. While in Canada and Mexico, the rate is just $2 per 24-hour period in which you utilize your plan. And in over 65+ partner countries, the rate is $10/day. If you don't utilize your plan in a 24-hour period, there is no charge. You do need to opt into the TravelPass option on your Verizon account to enable the feature – it is NOT automatic. This is however a super convenient and pretty affordable international option for occasional travel.

- **Monthly International Travel:** If you prefer a monthly plan, Verizon offers a variety of options. In Mexico and Canada you can add a monthly plan by line for $10 – $25/month that gives you a bucket of minutes (or access to 99 cent/minute rates), text messaging, and a monthly data allowance.

A special note for those protecting a grandfathered-in unlimited data plan. These older plans never had an international options, so adding on the Verizon international plans requires switching to a modern plan, which WILL discontinue your unlimited data plan! And you cannot get your unlimited data back after returning the States. Carefully consider this switch if your unlimited US data plan is important to you.

Crossing International Borders

T-Mobile
(http://www.t-mobile.com/optional-services/international.html)

T-Mobile is super attractive for international travelers.

All T-Mobile Simple Choice phone plans include free and unlimited international data (at 2G speeds) and text messaging in over 140 countries. And T-Mobile voice plans offer a 20-cent per minute flat rate for making and receiving calls.

With T-Mobile, your plan and device must be activated within the US before your trip begins, and your plan must be used "primarily" domestically. It is not intended for extended international travel.

If you'd like faster than 2G speeds, you can add on a data pass for $15/day for 100 MB 3G-speed data, or you can spend up to $50 for 500 MB of 3G-speed data, good for 14-days.

For Canada and Mexico, T-Mobile allows you to use your entire domestic plan on their roaming partners in a program called Mobile Without Borders. This means you get unlimited voice calls and text messaging, and access to all of your domestic high speed data allotment while traveling. If you have a T-Mobile Unlimited Plan, then you have unlimited data north or south of the border too.

For prepaid T-Mobile customers on the MetroPCS sub-brand, roaming is included in Mexico & Canada for just $5/month extra.

AT&T
(https://www.att.com/shop/wireless/international/roaming.html?tab=1)

AT&T offers the Passport plan add-on, for as low as $30/month, which offers packages of discounted international voice calls, unlimited texting, Wi-Fi hotspot usage, and a little bit of high speed cellular data in over 150 countries.

Passport plan prices start at $30/mo for 120MB data, or $120/mo for an 800MB package.

In early 2015, AT&T purchased two Mexican carriers in effort to create a unified North American network. Starting in late 2015 this effort began to bear fruit, and customers can add on 'Mexico Roaming Bonus' in their international settings at no charge – this enables free talk and text while in Mexico, and provides 1GB/mo of data usage (with overages at $20/GB).

Sprint
(http://support.sprint.com/support/international)

Sprint's international plans are a bit complicated to understand – there are two different programs that you can choose, they are free to sign up for, but you do need to sign up at least three days before your trip – and you need to pick the plan that best fits your needs.

Sprint's "Open World" plan includes free unlimited calling and text messaging in Canada, Mexico, and many other Central and South American countries – and you get 1GB of high-speed data to use internationally per month – with additional data billed at $30/GB.

You also get free texting and voice calls for 20-cents/minute in many other countries as well – with 2G speed data charged at $30/GB.

Or – you can sign up for "Sprint Global Roaming" and get free unlimited data, text messaging, and voice calls for 20-cents a minute – but data speeds are limited to 2G (64kbps) speed on the Global Roaming plan, even in Canada and Mexico.

While the 2G speed data may be unlimited - if you need faster access, you can purchase 3G speed data packages starting at $15 for 100MB for a day, up to $50 for 500 MB to use in 14 days.

Do keep in mind that Sprint does not have nearly as many international roaming partners as the other carriers - much of Asia, Africa, and Central Europe is not covered by either Sprint plan.

Google Project Fi
(https://fi.google.com/about/faq/#international-usage)

Google's "Project Fi" cellular service utilizes T-Mobile and Sprint behind the scenes domestically, but also offers a very flexible data usage package while traveling.

When in partnered countries (roughly the same 140+ that T-Mobile offers), you just pay the standard $10/GB rate that you pay in the US, except speeds are capped at 256 Kbps.

Going Native: Getting a Local SIM

If you're planning more extended time in a country, it may be worthwhile seeking out options with the local carriers to get a local prepaid or no-contract cellular plan.

Most of the rest of the world has standardized on GSM technology for their cellular networks – and if you have an unlocked phone, tablet, or hotspot that is GSM compatible (almost all are) you will be able to pick up a SIM card in many countries, and by putting it in your device you can make calls and surf the internet from your own tech at local rates.

All LTE phones activated on Verizon plans are all sold unlocked, but devices purchased from other carriers may be SIM locked and will refuse to work with foreign SIM cards. Before your trip you will want to request your device be unlocked, and assuming that your phone is fully paid for and a few other conditions are met, the carriers must abide.

Be sure to handle unlocking before you leave the country, as it can be difficult to do from afar.

Prepaid phones in particular can be difficult to get unlocked, especially for newer accounts.

It is also worthwhile to make sure your device can use the LTE frequency bands used by the cellular carriers where you are traveling. Many US devices are not compatible with the LTE bands used in much of the rest of the world, and will thus at best give you 3G speeds.

Here are some other options for global cellular services that won't require tracking down a new SIM in each country:

- iRoam.com sells a global SIM card that works worldwide. They charge 39 cents per megabyte outside the US (and 9 cents within). That equates to over $390 per GB. Not exactly an inexpensive way to go, but it does save a lot of hassle if you'll be moving through multiple countries.

- Telecom Square (http://mobilewifi.telecomsquare.us) — for a flat daily rate starting at $12.95/day, you can get access to unlimited data in a variety of international locations. This is even available to Canadian citizens traveling into the US.

Canada

Canada is a very large country, with a lot of wild unpopulated areas. Internet just simply isn't as abundant as one might hope for, and there are hurdles for US residents to get set up with their own independent connectivity source.

But with a little planning, you should be able to travel in Canada with at least moderate internet usage. Anyone depending on lots of internet for their mobility may find it more difficult and expensive while traveling northward than they would like. But it is getting easier.

If you're working online from the road and wanting to traverse through Canada, you may want to carefully consider the costs and logistics for mobile internet (as well as temporary work visa vs. business visitor issues — but that's another topic).

Sometimes the best strategy is to explore Canada while in more of a tourist mode without stressing about connectivity – especially if you are making the long drive up to Alaska, well away from the more populated border.

Wi-Fi

According to most RVers we've talked to who have ventured to Canada, Wi-Fi seems to be present and usable in most campgrounds and RV parks – and many utilize them for their primary internet needs.

Aside from campgrounds, Wi-Fi hotspots – like in most places across the world – are accessible from cafes, coffee shops, restaurants, libraries, breweries, and more.

Of course, public Wi-Fi hotspots will vary from free to paid, and the quality of connection will vary highly. Having Wi-Fi repeating gear on board can help improve the situation if the issue is range, but often the upstream connection is very limited – especially in small remote towns.

Cellular

Getting cellular service as a non-Canadian resident can be a bit tricky and expensive.

To get a prepaid plan directly with one of the carriers, most require a Canadian credit card or banking account, and as well as a Canadian address.

However many carriers have resellers, who will allow non-Canadian citizens to sign up with prepaid accounts – which tend to have low data usage amounts.

Once you have a cellular plan activated, service outside of metropolitan areas can be spotty and unreliable.

There's very little, if any, coverage in the wilds.

You'll also need to understand Canada's roaming areas, as unlike the US, different regions of Canada may be considered in or out of the local network for each of the carrier – depending on your plan.

There are three major carriers in Canada – Telus (www.telus.com), Rogers (www.rogers.com), and Bell (www.bell.ca). Each has its strengths and weaknesses in different parts of Canada, so be sure to check their coverage maps for your travel plans before investing much energy into trying to get a plan set up with one.

- **For Telus** – Telus has two options for getting an account – either a direct prepaid plan if you can find a kiosk who will sell you one without a local address. And their reseller account is also under the name Koodo which offers never-expiring prepaid data and minutes.

- **For Rogers** – US residents can purchase a prepaid plan through reseller Fido (www.fido.ca) or chatr (www.chatr.com). There are divisions owned by Rogers, available in many malls and service centers. You can obtain a SIM card for your existing device, or purchase a new device – and then a flexible monthly plan can be purchased that is billed as you need it. Fido tends to be better for more extended stays with higher data limits, and chatr better for shorter stays.

- **For Bell** – US residents can purchase plans through reseller Virgin Mobile Canada (www.virginmobile.ca) by visiting kiosks in several chain stores like Walmart or The Source.

Another option is having a friend who is a Canadian resident add on a noncontract SIM card to their plan and you can reimburse them. We've read reports of others doing this with great success.

If you're sticking to metropolitan areas, you may be able to hook up some very affordable and fast prepaid mobile internet deals with www.Mobilicity.ca and www.Windmobile.ca – they offer service just in the five largest Canadian cities of Toronto, Montreal, Calgary, Edmonton, and Vancouver.

Another interesting option is Similicious (www.similicious.com/) – who takes the frustration out of searching for and activating a SIM as a non-Canadian resident. They'll ship you a SIM card in advance from chatr (Rogers network) or have one waiting at your first destination.

Rates for phone, texting, and data range from $45 to $60/month with up to 2GB of data included.

The option of unlimited data through Telecom Square (http://mobilewifi.telecomsquare.us) might also be an option worth looking into at daily rates of $12.95.

Tip: The easiest of all option of all however is to just snag a T-Mobile plan while in the US, and use it while in Canada.

T-Mobile has roaming partnerships with all three of the major networks, and will roam onto them all. For as low as $35/month for a 6GB tablet plan, this can be a super easy and affordable route to keep connected in Canada.

Crossing International Borders

T-Mobile will roam onto all of the Canadian carriers at full speed and give you the same benefits as your domestic plan (other than Binge On unlimited video streaming). If you have a T-Mobile Unlimited phone data plan, you even get unlimited high-speed data while roaming! Their prepaid brand, MetroPCS, also includes the option for a small monthly fee.

If you have a Verizon tiered plan, you can also enable Verizon's TravelPass service and gain access to your domestic plan on an as needed basis for $2/day.

Sprint Open World will also will give you access to 1GB/mo of high-speed data, or Sprint Global Roaming will give you unlimited slow 2G speed data.

Tips for Canadians Traveling in the USA

For Canadians traveling in the US, here are some options worth considering:

- **US Prepaid Plans** – Obtaining a pre-paid plan with the major US carrier might be feasible, but tricky. The carriers may require a US based credit card and mailing address. Between using a prepaid credit card, or picking up refill cards in stores – and using the mailing address of a US friend or relative – you might be able to navigate this.

- **Rent/Lease Unlimited Plan** – For Canadians wanting lots of data, there are rental unlimited plans that can be obtained on a month-to-month basis on many of the major carriers. Since you are signing up with a reseller, they should be easier to work with as a non-US citizen. They can typically be found on eBay, checking our carrier guide online (www.rvmobileinternet.com/four-carriers) and our full member guide to unlimited data plans also covers the Verizon option.

- **KnowRoaming** (http://www.knowroaming.com/) – Toronto-based SIM sticker maker is offering an unlimited data package for $7.99/day that covers customers in 55 countries they have roaming agreements with. You purchase a SIM sticker for GSM based phones and tablets for $29.99 that auto-detects when you leave your home country, and then authorize the charge while abroad. They can also supply a local phone number. We were not able to confirm if the service allows for personal hotspot or tethering of the data, or if it is on device only.

- **Roam Like Home** (http://www.rogers.com/ consumer/wireless/travel) – Effective November 2014, Canadian carrier Rogers launched a service for their customers traveling to the US. For just $5/day (with a $50/month maximum charge), customers can access their plans in the US. So if a customer has an unlimited talk & text plan with a 6GB of shared monthly data – the entire plan can be used in the US too. Customers just need to text the word 'travel' to 222, to enroll in the plan and they are only billed for days they use the service.

- **Roam Mobility** (http://www.roammobility.com/) – Utilizing T-Mobile's network in the US, Roam Mobility supplies a SIM card to Canadian customers that offers a variety of options for daily phone & data, snowbird seasonal packages and data only packages. Options start as low as $3.95/day.

- **WIND Mobile** (https://www.windmobile.ca/) – Newer Canadian carrier WIND Mobile began offering unlimited US roaming in 2014 as part of its regular plans, which may be an option for some.

Satellite

Traveling across borders is where satellite internet has a definite advantage.

If you have a legacy HughesNet or Mobil Satellite plan and the equipment on your RV, you can subscribe to a satellite that covers part of Canada and be connected anywhere you have access to the southern sky.

There are no changes you need to make to your plan when you cross the border other than perhaps switching satellites, if needed. And the same rates and data limits you have stateside apply while across the border.

The RTC HughesNet K-band spot beam based plans however do not cover outside of the continental US.

Mexico

Traveling to our southern neighbor seems to be a bit more accessible for bandwidth junkies, as long as they are willing to deal with slower speeds than they're used to in the States.

We've heard of particular issues with obtaining enough fast internet for online working nomads in some of the more remote coastal areas like Baja.

Wi-Fi

Wi-Fi is available in some campgrounds, cafes, libraries, and public centers. Depending on the area, it may be slow and overloaded, particularly the farther from population centers you get.

Cellular

AT&T has made major strides in creating a North American cellular network by purchasing Mexican cellular provider Iusucell. Once it completes the integration, AT&T's stated goal is to provide 'one network, one customer experience.' – which may make connectivity to and from Mexico a breeze.

In November 2015, AT&T added Mexico Roaming Bonus as an international offer that gives free unlimited calling and text, and 1GB of data in Mexico with just $20/GB overage charges.

T-Mobile also now treats Mexico within its Mobile Without Borders, and allows customers to utilize their domestic plans while traveling south of the border. Their prepaid brand, MetroPCS, also includes the option for a small extra fee.

Verizon now includes Mexico in its TravelPass international program, allowing customers to access their domestic plan at just $2/day as they need it.

US citizens can also easily pick up a Mexican SIM card from one of the two major carriers – Telcel and Movistar. If you're picking up a phone service in Mexico, it's very important to note that the country is divided into nine regions, and calling is considered long distance between them. So if you pick up a phone service in Region 2, your number will be assigned there. If you travel to another region, your number will now be roaming in the new region for local calls. If you'll be making a lot of in-country calls, it may be best to pick up SIMs for each region you plan to travel.

For data, however, there is no roaming limits when it comes to regions.

Telcel (www.telcel.mx) is the largest carrier in Mexico and has the most coverage, especially outside of major metro areas. Speeds are reported to be at 3G in most places.

You'll need an unlocked GSM device (UMTA/HSDPA 850/1900MHz), such as a device off of AT&T or T-Mobile, to utilize Telcel service. Most recent Verizon phones are unlocked and will work too.

You can preorder a SIM card while in the US, and then just pop it in your device when you cross the border – or you can stop in any Telcel office, which are widely available countrywide. According to other travelers, you may have difficulty setting up a new account at a kiosk station, unless you speak Spanish well or happen upon someone overly helpful. The larger offices tend to keep an English speaker on staff to assist visitors to the country. The SIM card (called "chips" in some places) will cost about 149 pesos.

The Telcel Amigo service (www.telcel.com/portal/ personas/amigo/detalles/ internet_amigo.html? mid=1107) is no contract and pay as you go.

You put money on account either online (use Google translate to change the screens from Spanish to English), or at any "recarga amigo" station (located in many grocery stores, convenience stores, or gas stations). And then you order your plan by sending SMS codes (the "Clave" column on the above linked table) to "5050." You can check your balance by texting *133#.

If you're going to be in mainly metropolitan areas in Mexico, then Movistar (www.movistar.mx) may be an option for you. Movistar has less coverage throughout the country, but it is cheaper and faster. They have offices to set up your service in the markets they serve.

Satellite

Traveling across borders is where satellite internet has a definite advantage.

If you have a legacy HughesNet or Mobil Satellite plan and the equipment on your RV, you can subscribe to a satellite that covers part of Mexico and be connected anywhere you have access to the southern sky.

There are no changes you need to make to your plan when you cross the border other than perhaps switching satellites, if needed. And the same rates and data limits you have stateside apply while across the border.

The RTC HughesNet K-band spot beam based plans however do not cover outside of the continental US.

Sample Setups

Everyone's situation is going to be unique. There's no easy way to make recommendations for what you should build into your arsenal without really assessing your needs.

But we can make general recommendations for some common scenarios.

The next pages will go over some sample needs for mobile internet that we've encountered amongst fellow RVers we've talked to – and the basic setup that we'd recommend to them.

Obviously, your mileage may vary and be influenced by factors unique to you, such as any contract you currently have in place or any specific usage needs.

None of these recommendation should be taken as a shopping list.

And of course – technology is changing all the time, and new options are constantly becoming available. So always do your homework first!

Mobile Internet Advisor

If you would like some personalized private advising, we do offer Mobile Internet Advisor as a service. We'll conduct a comprehensive interview with you about your specific needs, do a little research on your behalf, have a private phone or video chat session, and then deliver a written report with our recommendations for you.

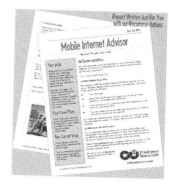

For more information, visit:
www.RVMobileInternet.com/advising

207

Lite Usage

Scenario: You just need to check email once or twice a day, do a little web browsing to research tomorrow's route and campground options, perhaps check into RVillage, and maybe an occasional video chat with loved ones back home. You're ok forgoing doing these things for a few days while you're off exploring a national park or out boondocking. Your internet needs are "nice to have" but not essential.

For this, having a tablet or smartphone device that can also hotspot to your primary laptop or computer may be sufficient. Get a plan with a smaller pot of data (maybe try 2–4 GB per month) to start, and scale as needed. You can supplement data from public Wi-Fi spots when it's available.

If you have multiple folks in your RV-hold and each of them wants his or her own cellular device, then we'd recommend a shared plan, if available on your carrier of choice. Or you may find it worthwhile to have multiple carriers within your household – such as one member on Verizon and another on AT&T.

Or, additionally, you may find that getting a MiFi-type mobile hotspot device added to your plan as a dedicated internet source may be desirable.

If you don't already have a preferred carrier in mind, we'd recommend Verizon as a starting place for most folks. Check their coverage maps, and see if they have service in the places you think you're most likely to go.

Moderate Usage

> **Scenario:** Keeping online is fairly important but not absolutely critical to your livelihood to be online all the time. But you'd really prefer to not go any considerable length of time without internet access.

You'll probably want an arsenal with at least two major connectivity sources, and those will depend on where you anticipate traveling, how much data you need, and your style of travel.

If you're planning to be mainly in populated areas with likely cellular signal and/or Wi-Fi hotspots – two different cellular options on different carriers may make sense, with public Wi-Fi being a backup and supplemental option.

If you don't already have contracts in place or preferred carriers and

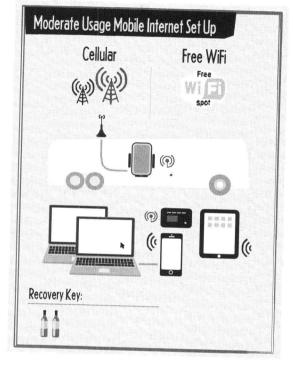

you'd like a decent bucket of data to use, we'd recommend getting a data plans on Verizon, AT&T and/or T-Mobile. With possibly one of them being unlimited.

You'll probably also benefit from a cradle-type cellular booster, like the weBoost Drive-4S. Or for boosting multiple devices, the 4G-M.

As always, Wi-Fi hotspots can help fill in the gaps. If you find yourself in areas with public Wi-Fi often enough, it may be worthwhile investing in a Wi-Fi boosting system, such as the WiFiRanger Elite or Sky.

If you're planning lots of remote boondocking away from cellular towers, you may want to replace our recommendation for a high data cap cellular plan with satellite internet instead.

High and / or Consistent Usage

Scenario: You need to be online fairly consistently, whether for work, school, or keeping in touch with family. You can probably survive being out of touch for a day or two.

You'd like a setup that keeps you connected most of the time, and you realize that there will be compromises to make.

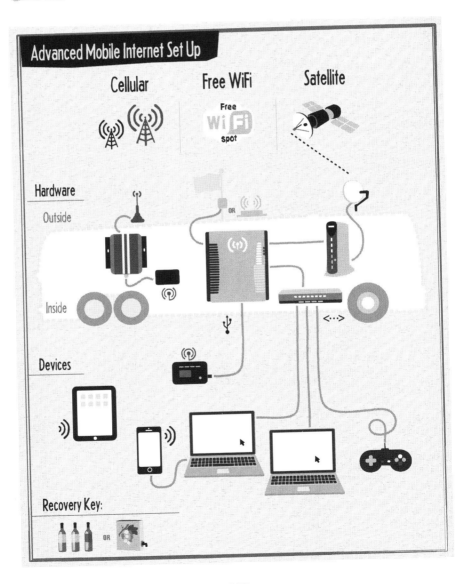

Sample Setups

You'll want to design around redundancy and having multiple backup options. You'll also want high cap and/or unlimited data sources.

- We recommend Verizon as the primary cellular network you utilize. The Verizon coverage map is pretty extensive, and the LTE very fast.

  - A Verizon Unlimited Data Plan will be your cherished friend. Consider obtaining one through assumption of liability, or even renting one from a trusted vendor.

  - Pay Verizon directly on a Mobile Share plan for the data you want (if you need more than 20GB/month, not recommended.)

- We recommend a data plan on a second network to help round out your cellular footprint. In our experience, AT&T is the best complementary network to Verizon. Between the two, if there's cellular signal in the area – you're likely to get online.

- If you stream a lot of video content or will be venturing into Canada or Mexico – a T-Mobile postpaid plan is also recommended to get Binge On unlimited video streaming, as well as an easy international data option.

- To increase your ability to utilize cellular in more places, a robust cellular boosting setup is almost essential.

- If you'll be staying in campgrounds with known reliable Wi-Fi, or driveway surfing with friends – a Wi-Fi setup to bring in public hotspots can be worthwhile. We recommend a router setup (WiFiRanger or Pepwave SOHO) and Wi-Fi antenna(s) on the roof of your RV. Remember, height is your friend – a mast of some sort with a directional antenna can vastly increase your range when stopped, and a more passive solution like the WiFiRanger Sky or Elite is great for shorter stops and closer-by hotspots.

- If you're planning travels outside of population centers where you may not have cellular signal or access to Wi-Fi – a satellite may be worthwhile. Carefully consider if the costs and logistics are worth it to you, or if you can juggle your desires to be out in the boonies with your connectivity needs.

- For times you need intense data, specifically seek out RV parks with reliable Wi-Fi networks and/or stay in monthly sites at parks where you can get cable or DSL installed to your site.

- Make space on board to carry lots of beer and wine to share with friends who have Wi-Fi to share and space to park your RV.

  This is our category and you can see our personal current arsenal later in this chapter.

Absolutely Need to Be Online

Scenario: You absolutely, positively, must be online and connected no matter where you roam, with the fastest possible speeds. You need to handle huge file transfers on a consistent basis, and can't go without streaming movies and TV shows. 500-1000 GB of a data a month would be great.

And you want to go everywhere – urban camping to boondocking in the wilds far from civilization.

Constant and fast internet access is critical to your mobile livelihood and happiness in life.

You're in luck, there is an ideal technomadic vehicle for you: Air Force One.

If you are traveling on any less of a budget, you need to come to grips with making trade-offs between speed, cost, convenience, mobility and coverage.

Your Author's Setup: Chris & Cherie's

As of the printing of this book (February 2016), this is the current setup that keeps us online most everywhere we go.

We're always tweaking this as new technology comes out, so be sure to check back with us here for any changes to our personal connectivity arsenal:

www.technomadia.com/internet

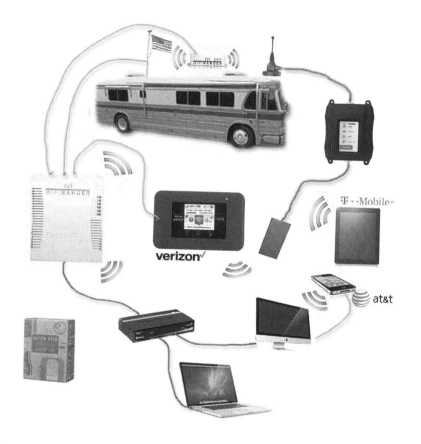

Cellular Options

To maximize our connectivity options, we keep a way online via all of the top three carriers:

- Netgear AC791L Mobile Hotspot with a grandfathered Verizon unlimited data plan ($69.99/month)

Sample Setups

- 2 iPhones on an AT&T Mobile Share Value plan with 40GB/mo shareable (via Personal Hotspot) data ($180/month – obtained under a 'double data' promotion)

- iPad Mini on a grandfathered AT&T unlimited data plan (not shareable) ($29.99/month)

- iPad Mini with a T-Mobile 6GB data plan, ($35/month)

Antennas & Boosting

- **Cellular: Booster & Antennas:** Our current primary booster is the weBoost Drive 4G-X. We use the included antenna, but we are currently testing several alternatives including a WirEng BoatAnt and Wideband Directional from weBoost.

- **Wi-Fi:** We have a WiFiRanger Go2 router inside paired with a WiFiRanger Sky2 on the roof. If we need longer range, we have a Ubiquity Nanostation we can put on a mast. We honestly don't use Wi-Fi all that often.

- **Mast:** When we're set up for a while somewhere, we put up a flagpole mounted to our hitch receiver. We not only displays our colors but also use directional antennas.

- If we need lots of bandwidth over a long period of time, we've been known to seek out a RV park were we can hook up to unlimited cable or DSL internet. And we've been known to head out to a library or cafe for Wi-Fi.

Router

We currently use the WiFiRanger Go2 as our RV-wide router that brings our cellular inputs and the occasional Wi-Fi into both wired ethernet and a private Wi-Fi network. We also use a generic gigabit ethernet switch to network our computers, NAS, and other gear together.

Fall Back Plan (LTTE Booster)

We keep a box of wine on hand for those times that nothing else works.

We've personal fans of the Bota Box RedVolution or Old Vine Zin. We call this our LTTE Booster – *Libation: Technology Tribulation Elimination.*

Your Author's Setup: Jack Mayer's

Be sure to check in with Jack for any updates to his setup:

www.jackdanmayer.com/communication.htm

Router

Our main router remains the WiFiRanger Go2. This brings together cellular and Wi-Fi, and enables me (Jack) to easily boost distant Wi-Fi with a WiFiRanger Mobile on the roof.

Other routers on hand are mainly used for testing, and include: Cradlepoint 95, 1000; Pepwave – all models of Pepwave back to six years ago…

I bring all my communications gear together in one location and most of it is wired for 12-volt use. My router runs all the time and we use it in motion. I also have a second WiFiRanger Go2 mounted in the truck in case we need it.

Wi-Fi Capture

A modified WiFiRanger Mobile is used most of the time, attached to the batwing TV antenna. This is modified to have an 8dBi omni antenna on it (like the WiFiRanger Marine or XT antenna upgrade).

I also have a WiFiRanger Sky2 on the roof that I sometimes use.

The batwing also mounts a Ubiquiti NanoStation M2 directional CPE for longer range Wi-Fi capture. I also have a dozen or more other Ubiquiti products on hand I use for various purposes. I've also tested a Wave WiFi Rogue Wave Wi-Fi capture device and occasionally use it – but my main capture is via the WiFiRanger.

I also have most of the Alpha products, Netgear range extenders, and at least half a dozen MikroTik devices that can act as an AP/Router/CPE.

I routinely mix and match all this gear to test various capabilities.

Cellular

Our primary devices are a Samsung Galaxy S4 and Samsung Note 5, both on a Verizon shared-data plan, and both hotspot capable.

We have our data "pool" set at 40 gigabytes via the Verizon double data plan offered in 2014 (we pay for 20 gigabytes).

A Jetpack AC791L mobile hotspot is used on Verizon, which also handles AWS XLTE spectrum. The Jetpack is attached to the WiFiRanger Go2 router wirelessly. I do not physically tether it via USB.

Cellular Boosting

At the moment we use a Wilson Sleek 4G (now known as a weBoost Drive 4G-S) as our primary booster.

Sometime in 2016 we will pick a new wireless boosting system that will better handle multiple devices.

Antennas

A selection of cellular and Wi-Fi omni antennas covers most of our needs.

Our primary cellular antenna is a WirEng BoatAnt omni. Our primary Wi-Fi antenna is either directional via a Ubiquiti NanoStation M2, or an 8 dbi omni mounted to a Ubiquiti Bullet or MikroTik device.

Full Time RVer Mobile Internet Setups

The range of setups out there is so diverse, and so individual to each household's needs and journey.

We keep an active list of full time RVers who have shared their mobile internet setups. You can check it out at:

http://www.rvmobileinternet.com/resources/other-full-time-rver-mobile-internet-setups/

In Our Testing Lab

We're constantly testing out gear so that we can assist our readers and premium members in picking out what will work best for their needs.

To see what we're currently playing with, and access reviews we already have available, check:

http://www.rvmobileinternet.com/lab

Wrapping Up: Top 10 Tips

There is a LOT of information in this handbook. To help wrap things up, here are some of our best words of wisdom distilled down to their core.

Tip #1: Embrace multiple pipes!

The more possible on ramps to the internet at your disposal, the more likely you are to find one that works. Embracing a diversity of connection types and networks is the best possible way that you can maximize your chances of getting at least somewhat of a workable connection, particularly since Wi-Fi alone is rarely going to be enough.

Tip #2: Soak up any (free) Wi-Fi you find!

Often the fastest, cheapest, and easiest way to get online is to use public Wi-Fi networks, and in some parts of the country and world these are growing increasingly easy to find. Many libraries, coffee shops, RV parks, motels, and even fast food restaurants now offer free Wi-Fi. Use it when you find it!!! But don't count on it – usably fast Wi-Fi is often a rare treat.

Tip #3: Understand roaming & coverage issues!

One place where all the carriers are a bit deceptive is around roaming. Because they want their networks to seem as large as possible, they go out of their way to hide from you that you may be roaming and running into usage limits hidden deep in the fine print of their contracts. Stay on guard, especially if you are on Sprint or T-Mobile!

Tip #4: Be aware near borders!

Beware of international borders! Most carriers charge an arm and a leg for international roaming (including onto the onboard cell networks offered on cruises now). If you are going to be anywhere close to an international border, make sure to turn off data roaming on all of your devices.

Tip #5: Know your caps!

Most fixed-location internet connections are unmetered, but mobile data is very commonly capped, and often comes with overage charges for excessive use. Save your big downloads for the days you have unlimited access.

Tip #6: Re-Assess Your Arsenal Annually!

This technology changes often. The carriers come out with more attractive plans or promotions, and it's important to look over your plan about annually to see if you're getting the best pricing and setup. Cellular devices (phones & hotspots) in particular should be upgraded at least every 2 years, to make sure they can connect to the latest frequencies.

Tip #7: Learn parallelizing & batching!

Things like email and syncing RSS readers work wonderfully in the background on a slow connection. But web surfing can feel painfully slow if every new page takes minutes to render.

To deal with this, parallelize your browsing using multiple tabs. Whenever you see a link you want to follow, select "Open Link in New Tab" and make sure your browser is configured to load tabs in the background.

Tip #8: Boost what you have!

A signal booster and/or an extensible antenna mast can work wonders to help you get online from afar. These systems aren't magical, but on several occasions they have made the difference between having a barely detectable signal and a barely usable one.

Tip #9: Stay safe out there!

The internet is a scary place – and public networks can be especially so. To keep yourself safe, never ever use the same password in multiple locations. To take your security and privacy even further, subscribe to a VPN service.

Tip #10: Final Tip – We Repeat – Manage Your Expectations!

If you are planning in advance on having good net days and bad net days (and even no net days), you can better manage your own expectations around what you will be able to get done online, and when. Managing your expectations is perhaps the ultimate key to avoiding frustration!

Glossary of Terms

Before you get frustrated wondering why you might need a POE to power your CPE to get remote 802.11g when you'd really rather have more dB on your LTE – check through this glossary, and soon it will all make sense.

Even for the terms you know, or think you know, checking the definitions here might take your understanding to the next level.

2G/3G/4G/ "True 4G" The "G" suffix has become a handy way to label various generations of cellular networks – though the official meaning has grown murky as carriers have taken liberties calling faster versions of third generation networking technologies 4G.

Technically, none of today's LTE networks actually even meet the official definition of "4G" until LTE-Advanced is deployed. The International Telecommunications Union has dubbed LTE-Advanced as "True 4G" – and snobby network engineers call today's LTE networks 3.9G.

Here are the technical standards and the marketing labels in use by the four major carriers:

Verizon: 2G (1xRTT), 3G (EVDO), 4G/LTE (LTE, XLTE)

AT&T: 2G (GSM/GPRS/EDGE), 3G (UMTS), 4G (HSPA+), 4G LTE (LTE)

Sprint: 2G (1xRTT), 3G (EVDO), 4G (LTE, Sprint Spark, LTE Plus)

T-Mobile: 2G (GSM/GPRS/EDGE), 3G (UMTS), 4G (HSPA+), 4G LTE (LTE, Wideband LTE)

5G

Fifth generation cellular networks are currently under active research and development – with some experimental trial deployments by Verizon, AT&T, and Google lately making the news.

The goal of 5G networks will be to enable ridiculously fast peak cellular data rates of over 10 Gbps, with network latency as low as 1ms.

This represents a 50x increase in network throughput and capacity compared to the fastest current 4G/LTE networks.

5G will enable a whole range of next generation applications ranging from autonomous connected vehicles to augmented reality to ultra-HD video streaming.

The actual technologies and frequency bands that will enable 5G have yet to be finalized by standards bodies – but 5G will likely take advantage of many chunks of spectrum ranging from long range UHF frequencies up through short range extremely high 60 GHz frequencies.

Though 5G technologies and trials will be increasingly in the news, the actual technical standards will not be finalized any sooner than 2019, with the first commercial networks deployed no sooner than 2020.

802.11

a/b/g/n/ac

The 802.11 standard is the formal name for the technology commonly referred to as Wi-Fi. The 802.11 standard comes in several flavors now:

- 802.11b – The first widespread standard. 2.4GHz, 11Mbps max.

- 802.11g – Faster. 2.4GHz, 54Mbps max.

- 802.11a – Operates on 5GHz, 54Mbps max

- 802.11n – Operates on 2.4GHz and optionally on 5GHz. By combining multiple MIMO antennas, speeds up to 600Mbps are possible.

802.11 **a/b/g/n/ac** **(continued)**	• 802.11ac – Operates on 5GHz and delivers blazing fast speeds over 1 Gbps over short distances, such as around a house. • 802.11ad – Also known as WiGig, intended to provide extremely high multi-Gbps speeds using high frequency 60GHz spectrum, with range limited to a single room. • 802.11ah – "HaLow" is the friendly name for the upcoming variant of Wi-Fi that uses 900MHz spectrum to double range, but HaLow is NOT intended for speed.
AirCard	The brand name for cellular modems from Sierra Wireless, but has become a generic term for modems that allows a computer to connect to a cellular network.
Amplifier	Pumps up the volume of a wireless signal. See also: "Booster" and "Repeater"
AMPS	Stands for Advanced Mobile Phone System, the original 1G analog cellular network widely deployed in the 1980s and 1990s. The AMPS network was phased out in the USA in 2008.
Analog	Analog radio signals are made up of waves that have not been digitized – meaning that anyone with a compatible tuner can listen in. Analog radio is easy to eavesdrop upon, is a very inefficient use of spectrum, and it suffers from static and other interference – especially as distance increases. The analog AMPS cellular network was at last fully shut down in 2008, but other wireless analog communication technologies remain – such as CB and ham radio.
Android	Google's smartphone operating system.
Antenna	An antenna takes electrical input and broadcasts it out as radio waves, or receives radio waves and provides an electrical signal out. An antenna needs to be designed and carefully tuned for the frequencies that it needs to support. An antenna optimized for traditional cellular networks may be useless receiving LTE signals. See also: "Frequency"

Glossary of Terms

Attenuation   Attenuation is the measurement of signal loss over distance in a wire or through the air, usually reported as dB/cm or dB/ft. This measurement is important for antenna cables – "low-loss" cables have much less attenuation over a given distance.

See also: "Gain"

Band (Radio)   The radio spectrum is divided up into bands based upon the signal frequency – ranging from the very low frequency bands used to communicate with submerged submarines to the extremely high frequency and beyond bands used for radio astronomy and exotic sensors and weapons.

Cellular networks all operate in the UHF (ultra-high frequency) band – which ranges from 300MHz to 3000MHz. Digital TV also operates in the UHF band.

Satellite communications and 5GHz Wi-Fi are in the SHF (super-high frequency) band, which is defined as 3GHz–30GHz.

Standardized batches of LTE frequencies are also known as LTE bands.

See also: "Spectrum" and "Frequency"

Bandwidth   All of us mobile internet junkies know that we crave bandwidth, but how many of us actually know what the technical definition of "bandwidth" is?

Any wireless broadcast actually extends across a range of frequencies – say, for example, the old analog TV channel 2, which spanned 54–60MHz.

The bandwidth is the size of that channel – in the case of analog TV in the USA, it was always 6MHz per channel. FM radio, in contrast, is 200kHz and AM radio 10kHz.

The larger the bandwidth allotted to a channel, the more information that the channel can carry. But with larger bandwidth channels set aside, there can be fewer of them without overlapping.

LTE supports a range of channel sizes – with bandwidths ranging from 1.4MHz to 20MHz wide.

(continued…)

222

Bandwidth

(continued)

In general, an LTE tower can support 200 active full-speed users per 5MHz of spectrum allocated. Once those 200 slots are filled, network speeds start to fall for everyone. This is why unlimited data is so hard to offer on cellular networks.

LTE-Advanced will support bandwidths up to 100MHz, accomplished by combing multiple separate channels together. This will allow for many more simultaneous users and much faster peak speeds.

Prime ranges of frequencies suitable for cellular data use are in very limited supply, and the major carriers have paid billions to buy up as much bandwidth as they can. Very few carriers own enough spectrum to offer 20MHz LTE channels in many places, but this is changing. 5MHz or less is actually much more common.

What Verizon calls XLTE is an example of a 20MHz LTE network.

This variation in available bandwidth is one of the reasons why not all LTE networks are created equal.

See also: "Frequency," "Refarming," and "XLTE"

Bandwidth Cap

Also know as a data cap, this is an imposed limit on the amount of data that can be used over a certain period of time. Internet service providers sometimes refer to their implementation of bandwidth caps as a "Fair Access Policy."

Bars

On mobile devices, bars are a visual representation of RSSI, the Received Signal Strength Indication – an arbitrary mapping of signal power received (in dBm) to a number.

This internally calculated number is usually then mapped to a simplified visual display – bars or dots.

Apple now has dots, most cellular devices have bars, and Wi-Fi devices come up with their own shapes all the time.

Also, bars are a great place to go to drown your sorrows when you can't get online any other way. As an alternative, consider keeping a box of wine handy.

Glossary of Terms

Bird-on-a-Wire (BOW)
Satellite internet dishes need to be precisely aimed at the satellite providing data service. Rather than require a second dish for television reception, a BOW mounts additional receivers precisely offset on the same dish being used for data to also allow that dish to pick up satellite TV service too.

This used to be commonly supported, but is rare now.

Bluetooth
Bluetooth is a slow, short range wireless communication technology most commonly used for wireless headsets and speakers. Though Bluetooth uses the same 2.4GHz frequencies as Wi-Fi, it is designed to avoid interference – and usually succeeds.

Booster
Cellular signal boosters are amplifiers used to improve signal reception. Typically they have connections for an inside and outside antenna, amplifying the signal from the external antenna and rebroadcasting it indoors.

bps (Speed Measurement)
Stands for bits per second. The smallest unit of computing is a bit – 0 or 1. Network speeds are measured in how many of these bits are capable of being sent per second. It takes 8 bits to make a "byte," and one byte usually represents a single character of text.

Broadband
In radio communications, broadband refers to a higher bandwidth transmission capable of transporting multiple signals simultaneously. This term became popular in the 1990s for describing any internet access technology faster than 56Kbps, which was the speed of a dial-up modem.

Broadband is relative term – and over time what it takes to be considered "broadband" has been gradually increasing.

As of early 2015 the FCC has redefined its official definition to require speeds of 25Mbps down and 3Mbps up for an internet provider to be able to claim that they are offering "broadband" service.

Cable Internet/ Modem
Internet service provided by a cable TV company, over the same physical network of wires that provides TV service. Cable internet can provide extremely fast service.

Glossary of Terms

Carrier

A cellular company is often referred to as a carrier because they carry transmissions on behalf of others. The legal status of being a "common carrier" means that cellular companies are not legally responsible for monitoring the content being passed over their networks. The liability for anything illegal falls upon the person doing the transmitting.

See also: "Operator"

Carrier Aggregation

One of the biggest challenges facing mobile network operators is that they have lots of 5MHz and 10MHz bandwidth chunks of spectrum, but to offer the fastest possible speeds LTE requires 20MHz, and LTE-Advanced can take advantage of up to 100MHz.

Carrier aggregation solves the problem by combining discontiguous channels from different chunks of spectrum to make a higher bandwidth virtual channel that can support faster speeds.

This is one of the core features of LTE-Advanced, and some carriers and devices have begun to support this technology.

See also: "LTE-Advanced," "Bandwidth"

CDMA

Code Division Multiple Access (CDMA) is the common name for the IS-95 and CDMA2000 standards for 2G and 3G networks embraced by Verizon and Sprint in the USA.

Unlike GSM, CDMA is not widely deployed around the world and is thus not well suited for international roaming. Most CDMA World Phones actually switch into a GSM mode while roaming.

Going forward, legacy CDMA networks are being replaced by LTE.

See also: "GSM"

Glossary of Terms

Cell/Cellular   A cellular network is based on dividing the coverage area into numerous slightly overlapping cells. A base station (aka cell tower) within each cell provides coverage to all the devices inside the cell and hands off responsibility to a neighboring cell when the user moves.

Cells can be as small as a room, or as large as 22+ miles (35+ km) in diameter. In congested or urban areas, smaller cells are used to allow for more people to use the network simultaneously.

Channel (Radio)   A channel is the combination of the frequency a signal is being broadcast upon and the bandwidth that it occupies.

In other words, the frequency is the address of the channel, and the bandwidth is the size of the house.

The more bandwidth a given channel takes up, the fewer of them you can have in a given slice of spectrum.

See also: "Bandwidth," "Frequency," and "Spectrum"

CPE   CPE stands for "customer premises equipment" and is the term used for commercial-grade Wi-Fi access points used by wireless service providers. Very often, the equipment actually providing Wi-Fi in a campground is a Ubiquiti or EnGenius brand CPE.

Advanced users can also use a CPE to maximize their own Wi-Fi range.

The WiFiRanger roof mounted Elite, Sky, and MobileTi, The Wirie AP+, and the Wave WiFi Rogue Wave products are all actually commercial-grade CPE's under the hood with user friendly simplified interfaces tacked on top.

Decibel (dB)   The logarithmic decibel scale is used to indicate the amount of gain an amplifier or antenna provides, or the loss introduced over wires. Positive numbers represent amplification, and negative numbers represent loss.

On the decibel scale, every 3dB represents a 2x increase, and every −3dB decrease represents 1/2. 10dB represents a 10x increase, 20dB represents 100x, 30dB represents 1000x, and so on.

The FCC-limited maximum gain for a mobile cellular boost is 50dB – this represents an increase in signal of 100,000 times.

Glossary of Terms

Decibel-Milliwatts (dBm)	It is very common to see signal strength expressed as a negative number with the units "dBm," which stands for decibel-milliwatts. But what do these numbers mean, and how should you compare them? And why are they negative?
	On the dBm scale, a value of zero indicates 1 milliwatt of power. The decibel is a logarithmic scale, so every 3dBm increase represents a doubling of power, and every 10dBm represents a 10x increase. So 3dBm represents 2mW (milliwatts), and 10dBm represents 10mW, and and 20dBm represents 100mW, and so on.
	The pattern works for negative numbers too. −3dBm is half a milliwatt, and −10dBm is 1/10th of a milliwatt.
	By the time a radio signal reaches a receiver, the amount of power left is incredibly small – and thus is measured in negative dBm. A great signal for a wireless device would be a −50dBm signal strength. A barely useable signal is −100dBm, which is 100,000 times weaker. And even the most sensitive receivers will have a hard time picking up a −110dBm cellular signal, which is a million times weaker than a −50dBm signal.
	It is amazing that such sensitive technology exists and can be packed into your pocket. But GPS radios are even more amazing – the typical signal received from a GPS satellite is around −127dBm!
DHCP	DHCP (Dynamic Host Configuration Protocol) is the magic that happens behind the scenes that allows your computer and other devices to automatically configure to connect to the upstream router. The router uses DHCP to assign each device an IP address and tells it what DNS service to use. If your computer and the DHCP server get out of sync, learning how to tell your computer to "Renew DHCP Lease" might be what it takes to fix things. (Rebooting accomplishes the same thing too.)
Dial-Up	Back in the dark ages, people used a telephone modem to dial a phone number of an ISP to connect to the internet.
Digital	Digital signals are made up of zeros and ones. All cellular networks today are digital networks. A digital network suffers drop-outs with distance, until eventually the signal is lost.
	See also: "Analog"
Dish	The saucer-shaped antenna that focuses signals from a satellite onto the low-noise block (LNB).

Glossary of Terms

DNS   DNS stands for "Domain Name Service" – This is the service that translates a name like "google.com" into an IP address like "206.181.8.251." Think of it like the white pages for the internet.

DSL   Stands for "Digital Subscriber Line" – home internet service provided via telephone wires, usually by a local phone company.

Ethernet   Wired networks are commonly called ethernet networks. Ethernet wires look like phone wires, though the jacks at the end are wider. The ethernet wire is sometimes called Cat-5.

FaceTime   Apple's proprietary ultra-simple video phone technology. With FaceTime, you can make a video call to other iPhone and iPad users, as well as to Mac laptops and desktops. But PCs and Android devices cannot make or receive FaceTime calls.

Frequency   Frequency is defined as the number of times a repeating event happens per second, and is measured in Hertz (or Hz).

A typical cellular radio might operated at 700MHz – which means that the peak of the radio wave passes 700 million times per second.

When it comes to radio waves, lower frequencies travel farther and can more easily pass through walls and obstructions. Higher frequencies are more easily blocked and are better suited for shorter range use. The advantage of higher frequencies is that there is much more bandwidth available – making it easier to increase network capacity and speeds.

To keep radios from broadcasting on top of each other, most of the radio spectrum is managed by the government and only licensed broadcasters can transmit on a ranges of frequencies that they "own."

The primary cellular frequency bands in use in the USA:
- 700MHz – AT&T and Verizon LTE.
- 800MHz – Cellular
- 1700/2100MHz – AWS
- 1900MHz – PCS

The primary Wi-Fi frequencies in use:
- 2.4GHz – 802.11b, 802.11g, 802.11n
- 5GHz – 802.11n, 802.11a, 802.11ac

See also: "Bandwidth," "Spectrum"

Glossary of Terms

Gain

The gain is the increase in signal provided by an amplifier or an antenna. Gain is logarithmic and reported in dB – a 3dB gain is a doubling in signal power. A 10dB gain is a 10x increase in signal power.

See also: "Attenuation"

Gbps

(Speed Measurement)

Gbps stands for gigabit per second and is also often written as Gb/s. A gigagbit is made up of 1,000 megabits.

Computers often have "Gigabit Ethernet" wired networking ports, but gigabit ethernet routers remain somewhat rare.

The new 802.11ac Wi-Fi standard paves the way for gigabit Wi-Fi over short distances, and the evolution of LTE into LTE-Advanced paves the way for gigabit cellular connections!

See also: "bps" and "Mbps"

Ground Plane

Many antennas are designed to need a ground plane to function properly, especially the common magnetic-mount antennas many cellular boosters come with.

A ground plane is created by providing a metal surface underneath the antenna. Steel works great (like a car roof), and the magnetic antennas stick beautifully to roofs like this. But aluminum also works as a ground plane – you just need to find a different way to hold the antenna secure to the nonmagnetic surface.

A rubber or fiberglass roof is useless as a ground plane, but you can instead use a small metal disk attached with adhesive to create one. A 3.5" diameter disk would be minimum, and 8" or more is better. Some people actually use a cookie sheet!

GSM

GSM stands for "Global System for Mobile Communications" and is an international standard for 2G cellular networks. In the USA, AT&T and T-Mobile built their networks on the GSM standard.

Central to the GSM standard is the SIM card, making it relatively easy to move service from one device to another. GSM phones are often able to roam onto other GSM networks around the world.

Going forward, legacy GSM networks are being replaced by LTE.

Glossary of Terms

Hangout Google's group video chat technology.

See also: "Skype" and "FaceTime"

Hotspot A hotspot is the common name for a Wi-Fi network. A public hotspot allows those nearby to connect without needing a password. A "personal hotspot" is the term often used to describe a private hotspot made by a smartphone or MiFi to allow other authorized devices to share a cellular connection and get online.

iOS The operating system inside Apple's iPhone and iPad.

IP Address Every site on the internet has an IP address. Behind the scenes, this numerical address is used to communicate instead of friendly names like technomadia.com.

See also: "DNS"

IPv6 The internet is running out of IP addresses to assign to newly connected devices. IPv6 is the next generation of networking protocols that pave the way for trillions of connected devices and beyond.

LTE is designed from the ground up based around IPv6.

ISP (aka Internet Service Provider) Who is providing your internet service?

That company is your Internet Service Provider (ISP).

Jetpack Verizon refers to mobile hotspot devices as Jetpacks, and the term is often used interchangeably with MiFi.

Kbps (Speed Measurement) Stands for kilobit per second and is also often written as kb/s. A kilobit is made up of 1000 bits, or 125 bytes (text characters). Old telephone modems were capable of receiving 56Kbps speeds – painfully slow by today's standards.

2G wireless networks are typically measured in Kbps.

See also: "bps" and "Mbps"

LAN LAN stands for Local Area Network, as opposed to WAN, which stands for Wide Area Network. Whether wired or wireless, the network inside your RV is considered a LAN.

Glossary of Terms

Latency

When it comes to networks, latency refers to how long it takes for a remote server to respond to a request. The latency is often referred to as "ping time" by speed-testing apps and is reported in milliseconds.

Think of it as follows: When you click a link, how long before you see a new page start to load? That time delay is latency.

The average latency for wired home internet services in around 30ms.

On a good LTE connection, ping times of 75ms to 100ms are common, and this feels plenty fast in use. HSPA+ 4G and 3G networks are slower – averaging around 120ms to 170ms.

Satellite networks, on the other hand, unavoidably have massive latency resulting from the signal's speed-of-light round-trip to geosynchronous orbit and back. The best satellite systems achieve a latency of 600ms, and over 1000ms (one full second!) is not unusual.

A high-latency connection may be capable of fast raw transfer speeds, but the connection will feel slow for interactive use because of all the delays. Latency is particularly painful for audio and video chat applications (where the half-second delay leads to talking on top of each other), and for typing into remote terminal sessions.

And online action games are a really bad idea without a low-latency connection – otherwise, you will very literally be dead before you know it!

LNB

On a satellite TV or internet system, the LNB is the extremely sensitive reception antenna mounted on a boom at the focal point of the dish.

LNB stands for low-noise block.

LTE

The dominant 4G technology is known as LTE, which stands for Long-Term Evolution. The LTE standard is gradually supplanting the previous cellular technology standards that came before it.

LTE-A (aka LTE-Advanced)

The LTE platform was designed to evolve to support faster and more advanced networks. The next major step in the evolution of LTE is being called LTE-A or LTE-Advanced, and the next step beyond moving closer to 5G has recently been dubbed LTE-Advanced Pro.

One of the primary advances coming into LTE-A is support for channel bandwidths up to 100MHz, 5x the current 20MHz maximum. To achieve this, LTE-A will support combining multiple separate physical channels into a single larger virtual channel.

The design goal for LTE-A is to enable cellular networks to support speeds up to 100Mbps for mobile users, and up to 1Gbps for stationary users.

Just a few years ago, these speeds would have been considered fantastical. Soon they will be available in pockets everywhere.

See also: "Bandwidth"

LTE Broadcast (aka LTE Multicast or LTE-B)

LTE Broadcast is a technology that allows for a video stream to be broadcast over LTE to multiple compatible simultaneous receivers – useful for live TV service and at sporting events.

Verizon and AT&T have shown off public demonstrations of LTE Broadcast services in 2015, and the technology may begin to go mainstream in 2016..

LTE Plus

Sprint originally called its triband LTE network "Sprint Spark", but in November 2015 to put more focus on new carrier aggregation and antenna beamforming technologies being deployed, Sprint rebranded its most advanced service areas and devices as "LTE Plus" – replacing the Spark branding.

LTE-U (aka LTE-LAA)

LTE-U is an experimental technology to broadcast LTE signals in unlicensed spectrum – allowing carriers to deploy super-fast short range LTE access points that supplement their regular licensed LTE frequencies with unlicensed 5GHz spectrum, co-existing with 5GHz Wi-Fi.

The technology is also sometimes referred to as LTE-LAA (License Assisted Access).

T-Mobile and Verizon have indicated strong interest in deploying this technology, once the Wi-Fi interoperability issues are worked out and the standard is finalized.

Glossary of Terms

MB/s (Speed Measurement)

MB/s stands for mega*bytes* per second – be very careful not to confuse this with Mbps (mega*bits* per second)!

One MB/s is 8Mbps – and even computer savvy people often screw up and use the wrong abbreviation.

Networks are always measured in bits per second – Kbps, Mbps, and Gbps. But computer interfaces like USB ports and peripherals like hard drives often have their speeds described in megabytes per second – MB/s.

A common USB 2.0 port maxes out at 60MB/s, USB 3.0 hits 625MB/s, and Apple's Thunderbolt port can transfer 1250MB/s.

Mbps (Speed Measurement)

Mbps stands for megabit per second and is also often written as Mb/s (not to be confused with MB/s). A megabit is made up of 1,000,000 bits, or 125,000 bytes (text characters).

Ethernet-wired networks used to operate at 10Mbps, and now 100Mbps "fast ethernet" is common.

3G and 4G wireless networks are typically measured in Mbps, as are Wi-Fi network speeds.

See also: "bps" and "Kbps"

MiFi

"MiFi" is Novatel's trademarked brand name for its line of mobile hotspots, but the term is often used generically to refer to any similar device.

MIMO

MIMO stands for multiple-input, multiple-output – or, in other words, using multiple antennas working together to increase data speeds. MIMO technology is central to the latest Wi-Fi and LTE cellular standards.

Mobile Hotspot

A mobile hotspot is a small device, usually battery powered, which combines a router with a cellular modem – allowing the user to share a cellular connection with other nearby Wi-Fi devices.

Modem

Modem is short for "modulate/demodulate" – the technical terms for transforming digital data to be transferred over a wire or a radio. The term "modem" was popular in the old dial-up days and is also commonly used to reference a cable modem or DSL modem.

MVNO   Mobile Virtual Network Operator – a company that offers cellular service, but which does not own its own cellular network. MVNOs lease capacity from the major carriers, and are often actually owned by the larger carrier as well.

Network   A collection of devices connected together is called a network.

See also: "LAN," "WAN," and "Cell"

Operator   Operator is another label often used to describe cellular companies, since they are operating a network on behalf of their customers.

The network operator owns or controls the licensed radio spectrum and the network infrastructure necessary to provide service over that spectrum.

Contrast this to an MVNO, which does not own the network but which leases service from a network operator.

See also: "Carrier," "MVNO"

Oscillation   If the outside antenna for a cellular booster picks up the signal from the inside antenna, oscillation happens; if the booster is properly designed, it will shut itself down.

This is the same phenomena as walking too close to a speaker with a microphone – leading to a howling screech.

The best way to avoid oscillation is to put as much distance as possible between the inside and outside booster antennas, and to keep the antennas pointed away from each other.

Ping   How much time elapses before receiving a response to a signal is the ping time.

See also: "Latency"

POE   Some networking equipment (such as the roof-mounted WiFiRanger units) are powered over the ethernet network wire

(aka Power Over Ethernet) instead of via a dedicated power cord and power supply. This makes wiring much simpler, since only a single wire needs to be run to the device. To get the power onto the ethernet wire, a POE injector is used to energize the wire. The WiFiRanger Go has a built in POE injector on one of its ethernet ports.

POTS   POTS is shorthand for Plain Old Telephone Service – the old way of getting online via dialing a modem on a regular wired phone line. Also known as a landline.

Glossary of Terms

Refarming

When cellular carriers shut down older networks, they free up frequency spectrum that they can then reuse to support newer technologies. This process is called refarming the network.

The downside of refarming is that devices based upon older technologies will get slower and less coverage, and eventually become useless. But newer technologies are much faster and more efficient, making things (eventually) better for everyone.

Repeater

(Wi-Fi & Cellular)

A booster amplifies a signal – including any background noise. In contrast, a repeater actually recreates and broadcasts a new signal without the noise. Cellular repeaters are rare since to rebroadcast (and not just amplify) a signal requires close cooperation with the carrier. But Wi-Fi range extending systems typically are repeaters.

Roaming

Roaming is when a cellular carrier has agreements with other networks to utilize their towers, helping the carrier provide connectivity to customers who are just passing through areas the carrier doesn't directly service.

Though customers are rarely charged directly for roaming anymore, behind the scenes roaming costs the guest carrier a substantial amount.

Router

A router acts as the hub of a local network. The WAN (wide area network) connection from the router provides the upstream connection from your local network to the internet.

Usually a router is required to allow multiple devices to share a single internet connection.

RSSI

RSSI stands for Received Signal Strength Indication – an arbitrary mapping of the power received by an antenna (in dBm) to a number.

How this number is calculated varies greatly from device to device.

This is usually simplified into a visual bars or dots display.

Satellite Satellites in orbit above the earth are invaluable for long-distance communications.

Some satellites are geosynchronous, which means they sit directly over a fixed point on the equator and are always in the exact same position in the sky. But because they are so far away, precise aiming with a dish-style antenna is essential.

Others satellites are deployed in "constellations" with multiple satellites in a lower orbit working together, with the goal being that at least one satellite in the constellation is always in view.

Because low earth orbit satellites are much closer and are in constant motion, aiming at them is not practical nor needed.

SIM SIM stands for Subscriber Identity Module. The tiny SIM card is mandated on GSM and LTE networks, and identifies you to the network.

Swapping your SIM into a new phone essentially moves your service and phone number to that new device.

SIMs come in a range of sizes – Full-Size is extinct, but Mini-SIM, Micro-SIM, and Nano-SIM are all in common circulation. Because the Mini-SIM was the standard for so long, that size is often referred to as "full sized" or "standard sized."

Using cutters and adaptors, you can transplant larger SIM cards into devices that have smaller SIM slots, and vice-versa.

SIM cards are often tricky to find but can usually be found behind the removable battery of some devices, or in small tray that can be ejected from the side of a phone with a pin.

Skype A cross-platform audio and video chat platform now owned by Microsoft. Skype is probably the most widely used video calling system.

See also: "FaceTime" and "Hangouts"

Smartphone Phones that are actually computers running a general purpose operating system that can be extended with applications are called smartphones. Phones lacking an extensible operating system are called feature phones. Apple's iOS (used in the iPhone) and Google's Android are the two dominant smartphone operating systems.

Glossary of Terms

SMS (aka texting, txt) SMS stands for Short Message Service – it is a standard technology for sending text messages of 160 characters or shorter directly between cellular devices.

Spectrum The electromagnetic spectrum is the range of all possible frequencies of radiation, ranging from extremely high-frequency gamma rays and X-rays through visible light to infrared to radio waves.

The radio spectrum is defined as 300GHz to 3kHz, and it is broken for convenience into bands.

See also: "Frequency," "Band"

Sprint Spark Sprint's original marketing term for its tri-band LTE network, now superseded by "LTE Plus". Spark devices use 800MHz frequencies for long range, 1900MHz for medium range, and 2500MHz for enhanced speeds in major urban areas.

Tethering Generally, tethering refers to using a USB cable to share a cellular device's connection with a laptop or router. This is the wired version of creating a personal hotspot.

Throttling Throttling is the act of intentionally slowing down a cellular data connection to run at a slower speed. Some carriers have moved away from charging overage charges when you hit your monthly data limits and instead now throttle users down to slower speeds.

In some cases, the throttling can be severe – turning a broadband LTE connection capable of 50Mbps into a snail-speed connection struggling to deliver 128Kbps.

"Unlimited" Data "Unlimited" is the holy grail of mobile data, but read the fine print and beware of throttling because there are often limits lurking..

VOIP VOIP stands for Voice Over Internet Protocol, and refers to technology and service providers that allow for traditional phone calls to be placed over the internet. VOIP calls can be placed either via specialized applications or by plugging a regular landline phone into a VOIP adaptor that connects to the internet.

237

VoLTE **(aka Voice over LTE)**	Right now, to make a voice phone call, cell phones switch their radios back to the old voice network to make the call. This is why some Verizon phones can't surf the web or get email while a call is underway, and AT&T phones drop back to 3G data speeds until the call is completed. Voice over LTE changes things up by sending your voice call as data over the 4G/LTE data network. This keeps your data connection running full speed and allows for HD Voice, which will let your voice calls to other VoLTE phones come through sounding more CD-quality than AM radio. VoLTE has a lot of advantages – and it opens the doors to the carriers eventually being able to fully retire their legacy voice networks – freeing up more space for fast data. As of early 2016, T-Mobile and Verizon support VoLTE on compatible phones nationwide, AT&T has deployed VoLTE in over 30 states, and Sprint has not yet announced VoLTE plans.
VPN **(aka Virtual Private Network)**	A VPN encrypts all the data coming to and from your computer, and sends it to a remote VPN server that then connects to the public internet on your behalf. This prevents anyone on your local network from being able to eavesdrop on you in any way – they can't even tell what sites you visit. If you are traveling internationally, you can use a VPN server so that you appear still to be connected from your home country. Sites that block international traffic (like Netflix streaming) will still work, and other sites will continue to default to your native language. A VPN will, however, introduce a small delay in all your network connections.
WAN	WAN stands for Wide Area Network, as opposed to LAN which stands for Local Area Network. Routers often have an ethernet port labeled WAN, and this is where you connect the upstream network if you have a cable or DSL modem.
Wideband LTE	T-Mobile is labeling the places where they have deployed extra large 15MHz or more bandwidth LTE channels as Wideband LTE, which should be comparable to what Verizon is calling XLTE. As of January 2016, T-Mobile claims that it has upgraded the bulk of its nationwide network to support Wideband LTE.

Glossary of Terms

Wi-Fi
Wi-Fi is the popular name for local area wireless networking technology. The technical standards that define Wi-Fi and which ensure interoperability are the 802.11 standard. Typical Wi-Fi networks can communicate several hundred feet.

WiFi-as-WAN
Some routers support using a Wi-Fi network as their upstream data connection, and this is called WiFi-as-WAN. With WiFi-as-WAN, you can use your router to connect to campground Wi-Fi while still maintaining a private local wired and wireless network.

Wi-Fi Calling (aka VoWiFi)
Some phones and carriers support making and receiving voice calls over a Wi-Fi connection as well as over a cellular connection.

T-Mobile in particular has made this a core feature on its network, though AT&T, Sprint, and more recently Verizon have also begun to roll this technology out on certain compatible devices.

WiMAX
WiMAX was a competing 4G networking standard that for a while was seen as a potential rival to LTE, and was Sprint's initial 4G network of choice.

LTE won out in the end, and all the major carriers have converged on LTE as their networking technology of the future.

Sprint has stopped expanding its WiMAX network and shut it down entirely in November of 2015 – freeing up more spectrum for LTE.

Some fixed-location wireless internet service providers (WISPs) still use WiMAX to provide service – it is very well suited to this sort of use.

WISP (aka Wireless ISP)
The term WISP (wireless internet service provider) has typically been attached to ISPs that deliver service to fixed locations via long-range wireless signals, often in rural areas where cable and DSL are not practical options.

A WISP providing service may have a transmitter on a nearby tower or mountain, and then install directional antennas towards it on the sides of customers' buildings. RVers in fixed locations can sometimes take advantage of a WISP for fast unlimited connectivity.

XLTE
Verizon's extended LTE network, now live in many major metro areas.

The Ongoing Conversation

Mobile Internet Aficionados: Our Premium Membership

If this book is the textbook..

...the MIA is the classroom.

Our premium membership is designed for those who depend on mobile internet to enable their roaming lifestyle, but don't have the time to keep on top of it all themselves.

We keep our Mobile Internet Aficionados (MIAs) 'in the know' about new developments in mobile internet, providing analysis of industry relevant news and how it impacts mobile RVers. We also create exclusive in-depth content just for our members, and host a member's Q&A forum.

We'll keep on top of the industry, alert you to anything that might impact you and answer your questions – so you can concentrate on what drives you!

This technology stuff changes often. It's almost guaranteed that as soon as we submit the manuscript to this book to be published, things will change. A new product will be released, a plan will be changed, or a major announcement will be made.

When it does, we'll scour news sources and forums and keep in touch with our industry contacts – and write an analysis on what it means to us mobile RVers.

Via this service, you'll have access to:

- Our exclusive member's forums – where you can ask us questions and get answers.

- Our exclusive membership newsletter – where we'll analyze mobile technology news and report to you what you need to know, as well as alert you to any upcoming changes that might affect you.

- Exclusive in-depth content released to members first – such as reviews, equipment testing results, and in-depth guides. All content on the site is always advertising free too for members.

- Access to exclusive interactive webinars.
- Access to private mobile internet advising sessions with us.
- *Free Updates to this book* in PDF format.

If this stuff is vital to your mobile livelihood, we invite you to consider joining us at:

www.RVMobileInternet.com/membership

And even if you don't join Mobile Internet Aficionados, do check the site from time to time – we will continue to post general public and free updates there that might impact decisions you make about mobile internet.

Save $5 on Membership

Our Mobile Internet Aficionados premium membership includes a copy of this book in PDF eBook format, and any updates that might be issued during your membership term.

To thank you for already having purchased this book separately, we offer you $5 off the price of a new annual membership to the MIA.

When joining at www.RVMobileInternet.com/membership, select the full Mobile Internet Aficionado package membership. At checkout use the below code to save $5:

Join the MIA

Save $5 off a new Annual Membership

Use Coupon Code:

IBoughtTMIH2016

Private Mobile Internet Advising Sessions

Ever wished you could just ask someone what you need to keep connected on the road? Well, now you can.

Everyone's needs are unique, and highly dependent upon how important internet is to you, your budget and your travel style. We'd love to help you navigate this stuff, and offer personal advising sessions so we can take the time to get to know you and your needs to make an informed recommendation.

We'll be happy to book a session with you to discuss your goals of keeping online while being mobile and help you assemble an ideal mobile internet arsenal custom designed just for you.

We start with having you fill out an assessment interview, schedule a phone or video session, and then follow up with a custom written report with our specific recommendations.

For more information or to schedule a session:

www.RVMobileInternet.com/advising

Keep in the know as this stuff changes... for news, further articles, resources, and updated editions of this book visit:

www.RVMobileInternet.com

Join our **FREE** e-mail newsletter for monthly updates on what's changed in the mobile internet landscape (we'll also alert you to new editions of this book coming out.):

http://eepurl.com/0KJG1

You are also welcome to join our free public Facebook group for discussions with other RVers (ourselves included) interested in this topic:

www.facebook.com/groups/rvinternet

62407908R00141

Made in the USA
Lexington, KY
06 April 2017